I0817480

GRISSOM
WHITE
CHAFFEE
APOLLO 1

THE APOLLO PHOTO ARCHIVE

APOLLO 1 IN PHOTOGRAPHS

THE APOLLO PHOTO ARCHIVE

APOLLO 1 IN PHOTOGRAPHS

J.L. PICKERING AND JOHN BISNEY

WITH ED HENGEVELD

4880 Lower Valley Road • Atglen, PA 19310

Other Schiffer books by the authors
Apollo 7 in Photographs: The Apollo Photo Archive
978-0-7643-7010-6

Library of Congress Control Number: 2025930665

Designed by Beth Oberholtzer
Cover design by Jack Chappell
Type set in Eurostile/Collier/Arno Pro

ISBN: 978-0-7643-7009-0
ePub: 978-1-5073-0583-6

Printed in China
10 9 8 7 6 5 4 3 2 1

Published by Schiffer Publishing, Ltd.
4880 Lower Valley Road
Atglen, PA 19310
Phone: (610) 593-1777; Fax: (610) 593-2002
Email: info@schifferbooks.com
Web: www.schifferbooks.com

Contents

Foreword

When I was a very young girl my father would sit with my brother, my mother, and me in our backyard and look up at the Moon. He would tell us about how he was going to ride in a rocket ship and maybe one day ride it all the way up to the Moon and, he hoped, walk on the Moon. I did not question what he wanted to do or why; after all, to me he was a rocket man and my loving father. I was used to being around astronauts and their kids, so it was just something they did—fly, fly, and fly again. At the end, my father never made it to the Moon, or even into space.

I grew up with the Space Race and eventually went to work for NASA at the Kennedy Space Center. I worked there for more than thirty years. I saw almost every Space Shuttle launch and numerous landings. Even after my personal tragedy, I have always supported the NASA mission. After all, we are explorers and have been since the beginning of the Earth's creation. I think it is important that we continue to explore, get back to the Moon, and, I hope, one day put humans on Mars. I continue to watch any launch that I can from my backyard or at the Kennedy Space Center when I can make it up that way.

Even after so many years, I can still see the smile on my father's face as he explained how he was going to the Moon one day and how proud he was to be an astronaut. He set his sight on the goal and worked hard to get there, all the way to the end. I now like to think that he is up there on the Moon looking back at us all on this great big Earth.

Sheryl L. Chaffee
Sebastian, Florida

Introduction

Project Apollo was the culmination of the US efforts to land astronauts on the Moon. It officially received its name in July 1960, although the federal government had been eying a moon-landing program for several years, primarily as the result of a natural evolution in technology. But Apollo took on additional urgency in 1961, after President John F. Kennedy set a nine-year deadline to land first in a Cold War competition with the Soviet Union.

The first six American astronauts flew solo during Project Mercury from 1961 to 1963. Mercury was assigned two broad missions by the government's civilian National Aeronautics and Space Administration (NASA): first, to investigate man's ability to simply survive and perform in the space environment, and second, to develop the basic technology and hardware for manned spaceflight programs to come. The following ten two-man Gemini missions from 1965 to 1966 developed the techniques critical for the Moon flights, including extravehicular activity, and rendezvous and docking with other spacecraft.

This book presents a photographic history of Apollo 1 (Apollo/Saturn 204), planned to be the first manned flight of the American Apollo program in 1967. Three astronauts would test the Block I versions of the command and service modules in Earth orbit, launched by a Saturn IB booster. On March 21, 1966, NASA announced that Mercury and Gemini veteran Gus Grissom would be the command pilot. The first US spacewalker, Ed White, would serve as senior pilot, while thirty-one-year-old rookie Roger Chaffee would be the pilot.

Jim McDivitt, Dave Scott, and "Rusty" Schweickart were named as the backup crew. A target launch date of February 21, 1967, was set. By December 1966, however, a planned second Block I flight had been canceled as unnecessary, so its crew—Wally Schirra, Donn Eisele, and Walt Cunningham—was reassigned as the backup astronauts for what had become popularly known by then as Apollo 1.

On August 19, 1966, in a meeting with Apollo Spacecraft Program office manager Joe Shea, the crew expressed concerns about the amount of flammable material such as nylon netting in the cabin, used for holding tools and equipment. At the end of the meeting, the astronauts gave Shea an alternate version of their crew portrait with their heads bowed and hands clasped in prayer.

When prime CSM contractor North American Aviation (NAA) shipped spacecraft CM-012 to Florida from its Downey, California, factory on August 26, 1966, there were more than 100 incomplete engineering changes, with 623 more made later. Grissom was so frustrated by the training-simulator engineers' inability to keep up with the changes that he hung a lemon on the Apollo Mission Simulator in Houston.

On January 27, 1967, the crew entered Apollo 1 at Launch Complex (LC)-34 for a "plugs-out" test, to see whether the spacecraft would operate normally on simulated internal power while detached from all ground cables.

A number of technical problems cropped up, including poor voice communications. Grissom asked, "How are we going to get to the Moon if we can't talk between two or three buildings?" A simulated countdown was conducted at 5:40 p.m. EST to try to fix the problem.

The crewmen were running through checklists when a voltage transient was recorded at 6:30 p.m. Ten seconds later, after Chaffee said, "Hey," followed by scuffling sounds,

Grissom reported a fire. Chaffee added, "Fire! We've got a fire in the cockpit." Seventeen seconds after the first indication of a fire, the cabin ruptured. Expanding gases from the fire had overpressurized the CM.

Flames and gases rushed out into two levels of the pad service structure surrounding the CSM. Intense heat, dense smoke, and the use of gas masks designed for toxic fumes instead of heavy smoke hampered rescue efforts. It took five minutes to open all three hatches: a protective outer hatch cover for launch; an outer hatch, part of the spacecraft's heat shield; and a removable inner hatch.

After the inner and outer hatches were in place, the cabin air had been replaced with pressurized oxygen, and flammable material in the pure-oxygen environment allowed the flames to spread quickly. The fire had partly melted areas of the outer layer of Grissom's and White's nylon space suits. Grissom had removed his restraints and was lying on the floor of the cabin. White's restraints were burned through; he was found just below the inner hatch. He had tried to open it but was not able to do so against the internal pressure. Chaffee was found strapped in his seat, from where he was to have maintained communications.

NASA immediately convened the Apollo 204 Review Board. Although investigators never conclusively identified the exact ignition source, they attributed the fire to a wide range of design and construction flaws. The most likely cause was a spark from a short circuit in a wire bundle near Grissom's feet. After a thorough disassembly and inspection, the Apollo 1 CM was shipped in pieces to Langley Research Center in Hampton, Virginia, for long-term storage.

The Apollo program was delayed for twenty months while major safety upgrades were made, including developing a new combination hatch that could be opened quickly, replacing much of the flammable material, and switching to a nitrogen-oxygen atmosphere on the ground. A new flame-resistant, fiberglass-based fabric of woven Teflon-covered silica fiber—Beta cloth—was developed for the outer layer of the Apollo space suits, as well as for other in-flight uses.

At the end of their 1969 moonwalk, Apollo 11's Neil Armstrong and Buzz Aldrin left an embroidered Apollo 1 patch on the lunar surface. The names Grissom, White, and Chaffee were etched on a metal plaque also left on the Moon by the Apollo 15 astronauts in 1971. The plaque, which listed fourteen fallen astronauts and cosmonauts, was accompanied by a 3-inch sculpture representing them.

In 1970, the International Astronomical Union named craters on the far side of the Moon for Grissom, White, and Chaffee, and in 2004, NASA honored the Apollo 1 crew by dedicating three hills to the astronauts near the landing site of its *Spirit* Mars exploration rover. In 2017, Apollo 1's three hatches were put on permanent display at the KSC Visitor Complex on the sixtieth anniversary of the accident.

Acknowledgments

Matthew Beddingfield
Shane Bell
Howard Benedict
Andrew Chaikin
Herbert Desind
Mark Gray
Ken Havekotte
Juliana Shepard Jenkins
Alan Lawrie
Dave Pequet
Scott Schneeweis
Myra Taub Specht
Ron Specht
John Uri
Mark Usciak
Thomas Usciak
Ron Woods

CHAPTER 1

The Astronauts

The first manned mission planned for the Apollo program also produced a first tragedy: no US astronaut had previously died aboard a spacecraft. And although Apollo 1 is primarily remembered for that launchpad fire that claimed three lives, it would have established a number of firsts: the first flight of a three-person US spacecraft, the first manned launch of the most powerful booster at the time (the Saturn IB), the first time an astronaut had made three space flights (Grissom), and the first TV broadcast from a US spacecraft.

The mission was scheduled to begin on February 21, 1967, at 10:00 a.m. EST, with a five-and-a-half-hour launch window, and was planned to possibly last nearly two weeks, which would have broken the world spaceflight endurance record set by Gemini VII by about fifteen minutes. The crewmen would check out all spacecraft systems, including its main propulsion system engine. Two engine firings on the second day would raise and circularize their orbit to about 160 miles high. Nine medical, scientific, and technical experiments were also in their flight plan. Splashdown was planned for north of Puerto Rico in the Atlantic on March 7.

The shocking loss of life was a huge wake-up call for everyone involved in the program, and the resulting modifications to hardware and procedures were critical to the success of the American Moon-landing effort just two and a half years later.

Portrait of US Air Force (USAF) major Gus Grissom by National Aeronautics and Space Administration (NASA) photographer John Holland in the fall of 1964, taken in his studio at the Manned Spacecraft Center (MSC) in Houston, Texas

The Apollo I crew emblem is designed by Allen Stevens, a graphic artist with Apollo spacecraft prime contractor North American Aviation (NAA).

USAF major Ed White's formal portrait by Holland at MSC in 1964. Holland often inserts white paper shirt cuffs into the subject's jacket sleeves for his portraits. He holds a copy of *Aviation Week & Space Technology magazine*, as does Chaffee.

Holland's 1964 portrait of US Navy lieutenant commander Roger Chaffee at MSC

Grissom's first portrait as an astronaut in 1959

NASA selected seven military officers as its first group of astronauts in 1959 for Project Mercury, but with the US soon gearing up for a manned moon-landing effort, more crewmen were needed. By 1966, the space agency had more than fifty astronauts who had either flown during its next program, Gemini, or were training for Apollo. Each group officially carried a sequential group number but also had a nickname, beginning with Group 1, dubbed the Mercury Seven. Astronauts from the first five groups flew during the Apollo program.

Grissom is one of the first seven astronauts, posing in front of a Mercury full-scale model shortly after their selection while visiting McDonnell Aircraft Corp. in St. Louis, Missouri, on May 12, 1959. McDonnell is prime contractor for the Mercury and Gemini spacecraft. *Left to right*: Scott Carpenter, Gordon Cooper, John Glenn, Grissom, Wally Schirra, Alan Shepard, and Deke Slayton.

Grissom poses by the Multiple Axis Space Test Inertia Facility (MASTIF) at Lewis Research Center in Cleveland, Ohio, on July 10, 1960. The trainer, also called "the gimbal rig," simulates tumbling situations that a pilot might encounter during spaceflight.

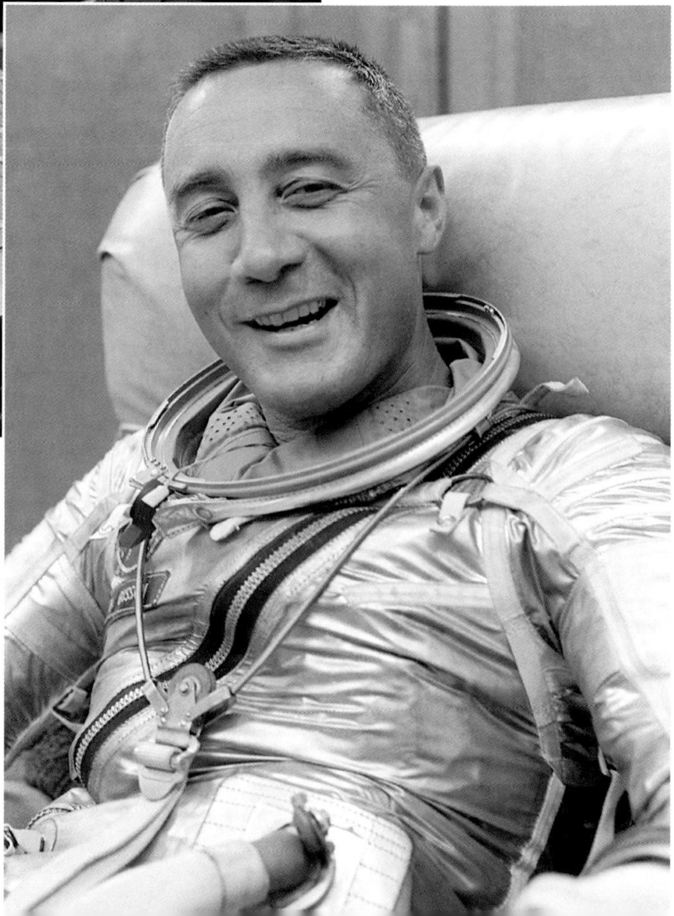

Grissom prepares for his launch aboard Mercury-Redstone (MR)-4 in the early morning of July 21, 1961, in Hangar S at Cape Canaveral Missile Test Annex, Florida.

MR-4 lifts off from Launch Complex (LC)-5 on a fifteen-minute suborbital flight, making Grissom the second American in space. The yellow mobile aerial tower, or "cherry picker," is retracted at right.

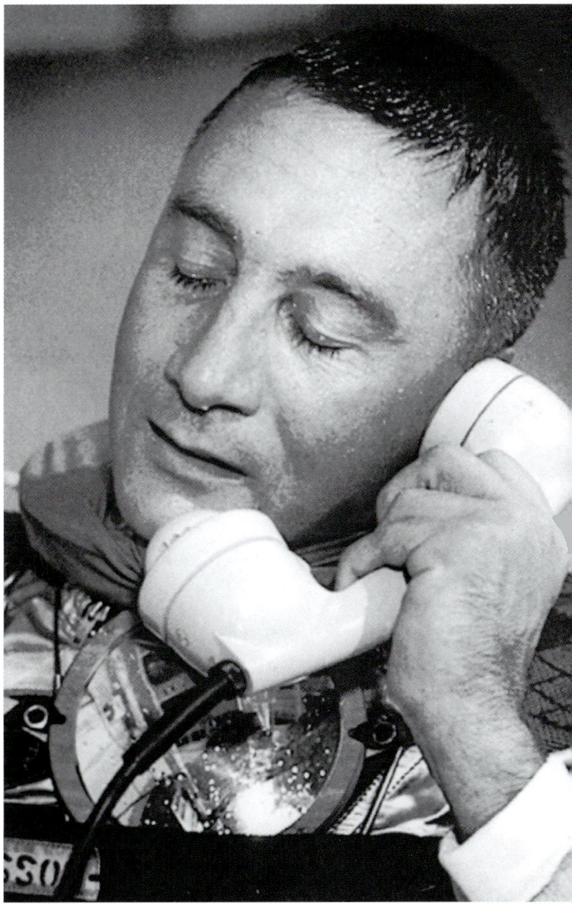

Grissom receives congratulations from President John F. Kennedy aboard the aircraft carrier USS *Randolph*.

Grissom rides a mule during geology training for anticipated lunar exploration at the Grand Canyon in Arizona. Ten astronauts from the first two astronaut groups take part from March 12 to 13, 1964, led by scientists from the US Geological Survey.

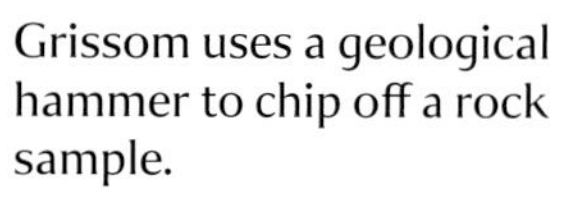

Grissom uses a geological hammer to chip off a rock sample.

Grissom poses in front of the Gemini Mission Simulator at McDonnell in St. Louis, Missouri, on May 22, 1964, while training as command pilot for the Gemini 3 mission. He wears a G2C pressure suit, an early version made by the David Clark Co. of Worcester, Massachusetts. During the flight, he and pilot John Young would wear an improved G3C model. *Bill Taub archive*

A Titan II booster launches Gemini 3 into orbit from LC-19 at Cape Kennedy Air Force Station (AFS) at 9:24 a.m. EST on March 23, 1965.

Clouds obscure the southeastern coast of Africa in this photo taken by Young from his right-hand window during the mission's second orbit.

Grissom (*right*) and Young speak at the Cape Kennedy AFS Skid Strip on March 25, 1965, two days after their three-orbit flight. They had arrived from recovery ship USS *Intrepid* and would next undergo a medical exam. "We're pretty well talked out after the past few days," Grissom tells reporters, but he promises a news conference that evening after a parade through Cocoa Beach, Florida.

White's first astronaut portrait in 1962

White is chosen for the second group of NASA astronauts, dubbed the New Nine. They pose with the models of Mercury, Apollo, and Gemini spacecraft after a news conference at the University of Houston's Cullen Auditorium on September 17, 1962. *Back row, left to right*: Elliot See, Jim McDivitt, Jim Lovell, White, and Tom Stafford. *Front row, left to right*: Pete Conrad, Frank Borman, Neil Armstrong, and John Young. Three of these men would eventually walk on the Moon; three others would circle it.

White visits NASA's Langley Research Center in Hampton, Virginia, in December 1962. *Bill Taub archive*

During his stay, White practices procedures in the Rendezvous and Docking Simulator in the background for Gemini spacecraft to dock with an unmanned Agena target vehicle. *Bill Taub archive*

White is among the Group 2 astronauts who undergo training in desert survival techniques east of Stead Air Force Base (AFB) at Carson Sink north of Reno, Nevada, from August 5 to 10, 1963. His head covering is made from a parachute.

White examines an outcropping during geology training at the Grand Canyon, Arizona on March 12, 1964.

White is pulled during water survival training at Naval Air Station Pensacola, Florida, in September 1963 as part of early Gemini training. *Bill Taub archive*

The prime and backup crews for the Gemini 4 mission pose with NASA's first photographer, Bill Taub, on the East Front of the US Capitol during a visit to meet members of Congress on April 29, 1965. *Left to right*: backup command pilot Frank Borman, prime command pilot Jim McDivitt, Taub, prime pilot Ed White, and backup pilot Jim Lovell. *Bill Taub archive*

White has closed his helmet visor and is ready to head to LC-19 for launch aboard Gemini 4 on June 3, 1965.

White floats outside Gemini 4 at the end of a 25-foot tether in this 16 mm still frame during the first US extravehicular activity (EVA) on June 3, 1965. He uses a Hand-Held Maneuvering Unit with a still camera mounted to it.

Gemini 4 is reflected in White's gold-plated visor in this photo by McDivitt, using a Hasselblad 70 mm camera.

Chaffee's first astronaut portrait in 1963

Chaffee is among NASA's third group of astronauts, known as the Fourteen. They pose at the University of Houston's Cullen Auditorium on February 5, 1964, two days after first reporting to MSC to begin training. *Front row, left to right:* Chaffee, Gene Cernan, Alan Bean, Charles Bassett, Bill Anders, Edwin "Buzz" Aldrin, and Russell "Rusty" Schweickart. *Back row, left to right*: Dick Gordon, Ted Freeman, Donn Eisle, Mike Collins, Walt Cunningham, Clifton "C. C." Williams, and Dave Scott. Of Group 3, four would eventually walk on the Moon; three others would circle it.

Chaffee rides a mule at the Grand Canyon during the first geological field exercise for the Group 3 astronauts on March 6, 1964.

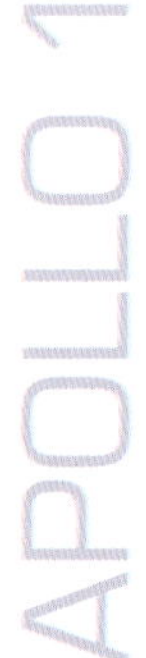

Chaffee (*lower left*) compares notes with instructor Dale Jackson (*center*) of the US Geological Survey during training at Sunset Crater near Flagstaff, Arizona, on May 1, 1964. Astronauts Mike Collins (*right*) and Dave Scott listen.

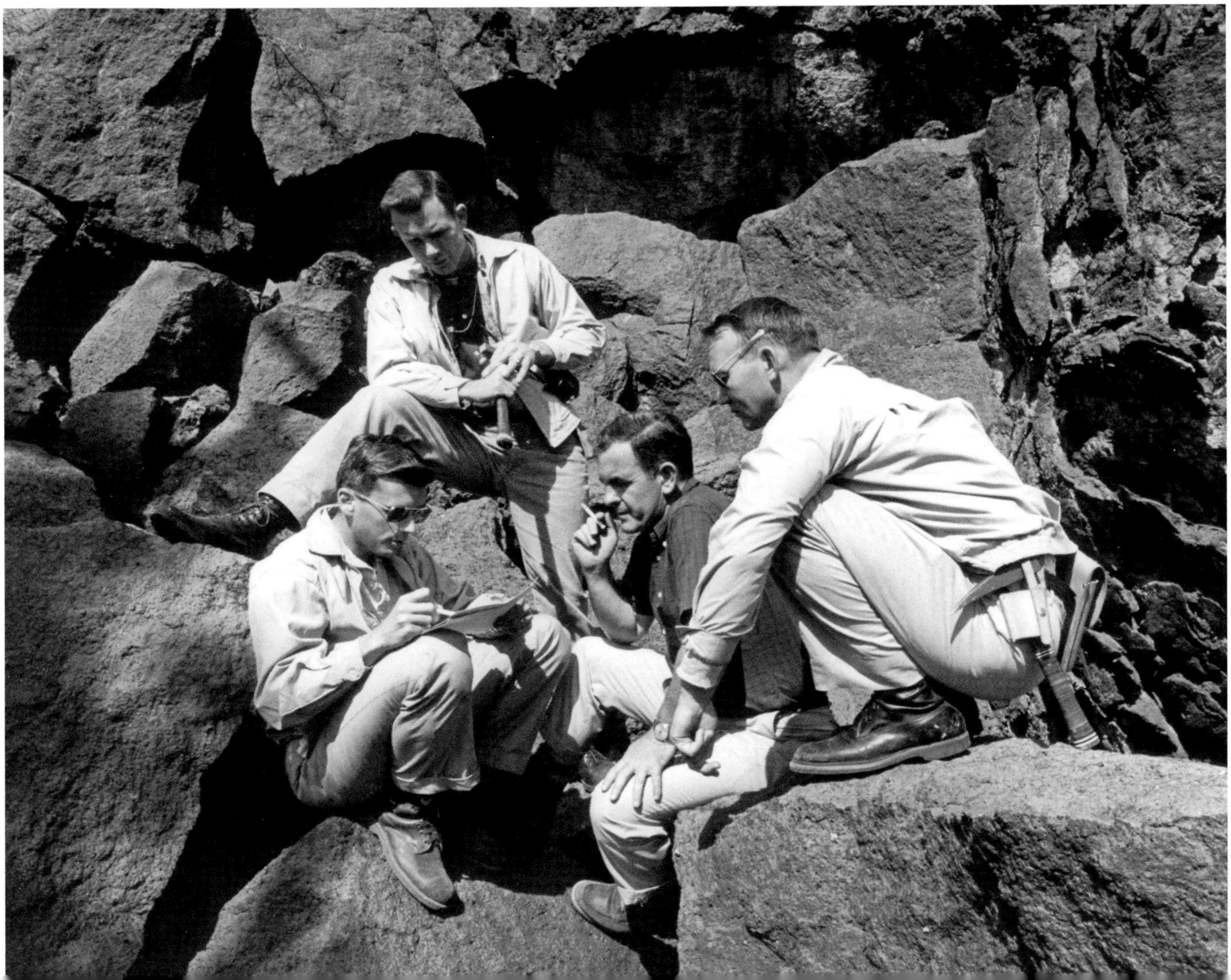

Chaffee takes in the view from the spacecraft access level of the LC-37 service structure on April 20, 1964, during a tour for the Group 3 astronauts of Cape Kennedy AFS. The sixth Saturn I (and the second from LC-37) is launched just more than a month later.

The first six American astronauts were solo pilots of their small Mercury capsules in low earth orbit. A one-part spacecraft with a reentry rocket package strapped to its heat shield, its interior volume was 60 cubic feet.

Mercury's successor, the two-man Gemini spacecraft, was also built by McDonnell Aircraft Corp. of St. Louis, Missouri. With an interior volume of 90 cubic feet, it provided slightly less room per astronaut than Mercury but was easier to ingress and egress with two hinged doors. As Apollo would, Gemini carried a separate adapter (service) module jettisoned before the manned module reentered the atmosphere and splashed down. This interim ten-flight program achieved the necessary skills for Apollo, including EVA and rendezvous and docking.

The Apollo spacecraft consisted of a three-man command module and an attached service module, which included a large engine used primarily to enter and leave lunar orbit. Its interior volume was 452 cubic feet. It could dock and undock with a separate lunar module built by Grumman Aerospace Corp., which accomplished the landings. Apollo was built by NAA under a NASA contract awarded in December 1961.

Chaffee wears garb fashioned from parachute fabric during desert survival training at Carson Sink east of Stead AFB from August 10 to 14, 1964. They spend three days and nights in the desert and receive classroom training.

Chaffee (*left*) and fellow Group 3 astronaut Gene Cernan take part in geology training at Katmai National Park in southern Alaska from June 29 to July 2, 1965. The area has a rich volcanic history.

Chaffee examines a sample during geology training in northern Iceland from July 12 to 16, 1965. The astronauts examined rocks found in glacial outwash channels that resembled the complexities of the lunar surface debris layer, and newly formed volcanic rocks.

The Apollo 1 crew portrait of (*left to right*) White, Grissom, and Chaffee, taken by MSC photographer John Holland on April 1, 1966

The Apollo 1 backup crew of (*left to right*) Scott, McDivitt, and Schweickart stands behind the prime crew.

Holland also shoots several gag crew photos, including a sadly prophetic image of the astronauts praying. Grissom gives a copy to Apollo Spacecraft Program manager Joe Shea on August 19, 1966, signed by the three with the inscription, "It isn't that we don't trust you, Joe, but this time we've decided to go over your head."

The prime crew has fun with an Apollo command module (CM) model and escape tower.

The backup crew plays with the model.

CHAPTER 2

January 1965–May 1966

S-C 012 S/M

S-C 012 C/M

North American Aviation (NAA) technicians work on what would become Apollo I, command module (CM) no. 012 (*left*), in Building 290 at NAA's Space and Information Systems Division in Downey, California, in January 1965. The service module (SM) is in a workstand (*right*). Spacecraft fabrication began in August 1964, and the basic structure will be completed in September 1965.

The crew compartment is in a test stand above its aft heat shield at Downey on June 23, 1965. Building 290 was the largest clean-room facility in the world when built in 1963, and has a high bay and a low bay. Final assembly and installation of subsystems into the CM are finished by March 1966.

From 1937 to 1967, North American Aviation designed and manufactured many significant US military aircraft and civilian spacecraft. In 1948, NAA built a new plant at Downey, California. In November 1961, NASA selected the company as the prime contractor for the Apollo command/service module (CSM), and later, the second stage (S-II) of the Saturn V booster.

NAA had proposed a 60 percent oxygen / 40 percent nitrogen mixture for Apollo's crew cabin, but NASA decided that a pure-oxygen atmosphere would be safer, less complicated, and lighter in weight. It was blamed in part, however, for the Apollo I fire, and relations between NASA and the company deteriorated over which was responsible.

In September 1967, NAA merged with Rockwell-Standard to form North American Rockwell (NAR), which later became Rockwell International, the prime contractor for the space shuttle orbiter.

The CM consisted of an inner structure (the pressurized crew compartment) and an outer structure (a three-section heat shield). The pressure shell was a sandwich of a welded-aluminum inner skin, an adhesively bonded aluminum honeycomb core, and an outer face sheet. The outer ablative heat shield was composed of a fiberglass honeycomb shell filled with phenolic epoxy resin, and consisted of the forward, crew compartment, and aft heat shields.

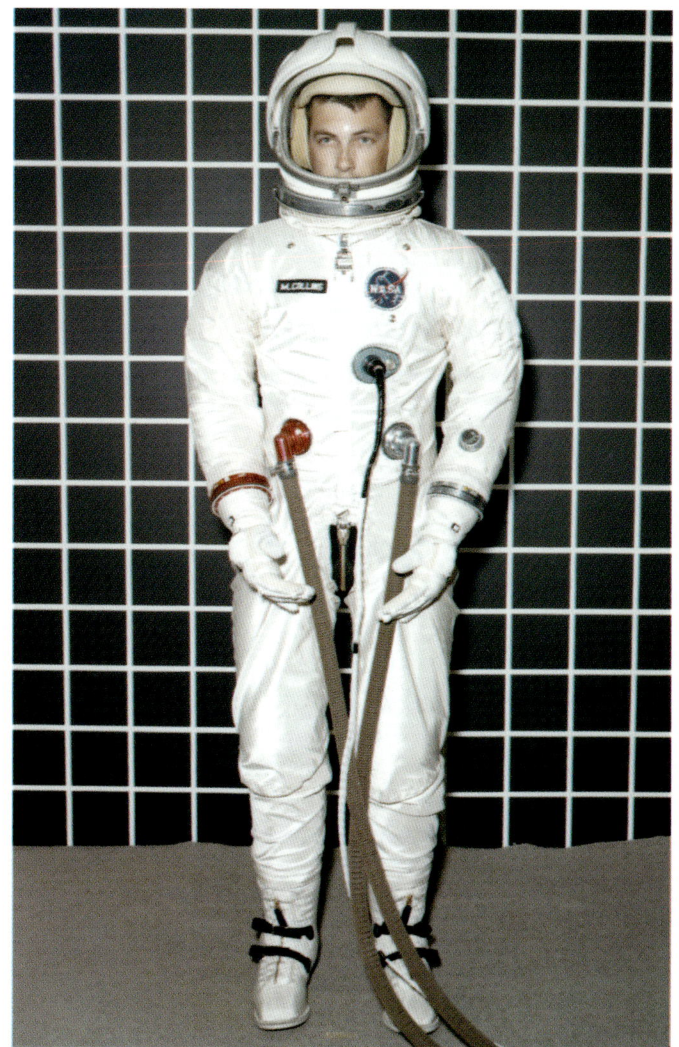

A test subject models an A1C space suit tailored for astronaut Mike Collins to use during training in July. It is a modification of the Gemini G3C and G4C space suits, manufactured by the David Clark Co. of Worcester, Massachusetts. The suits were chosen for the Block I Apollo missions, while a design competition is underway for a new Block II lunar EVA suit, primarily among David Clark, the International Latex Corp. (ILC) of Dover, Delaware, and Hamilton Standard of Windsor Locks, Connecticut.

Technicians at the Michoud Assembly Facility in New Orleans, Louisiana, prepare the first stage (S-IB-4) of Apollo 1's Saturn IB booster for final buildup in the fall of 1965; its heat shield has not yet been installed. Built by the Space Division of the Chrysler Corp., the stage would be transported by barge on December 7 to the Marshall Space Flight Center (MSFC) in Huntsville, Alabama, for static testing. The eight Rocketdyne H-1 engines, burning refined kerosene (RP-1) and liquid oxygen (LOX), would produce a total of 1.6 million pounds of thrust.

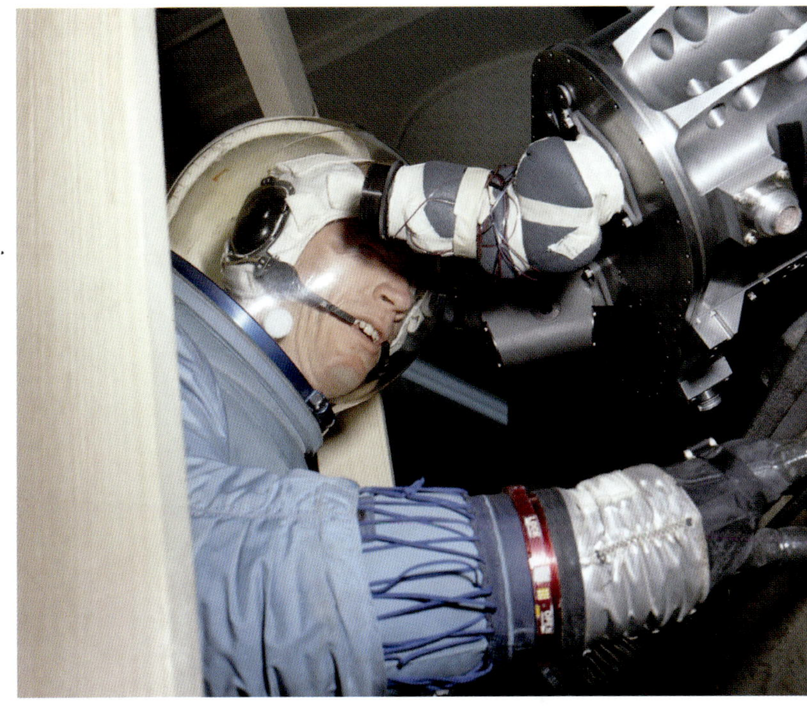

Astronaut Russell "Rusty" Schweickart wears an AIC pressure suit at the MIT Instrumentation Laboratory in Cambridge, Massachusetts, on January 14, 1966, during checkout of the optics for the Apollo Primary Guidance, Navigation, and Control System. The system was developed by MIT under the direction of Charles Draper, and the lab would later be named for him.

Schweickart wears a prototype of the Apollo A7L pressure suit without its white outer layer at the MIT Lab on January 14. The Apollo 1 crew is officially announced two months later, with Schweickart on the backup crew.

Left to right: Astronauts Grissom, Walt Cunningham, and Schweickart pose in front of Apollo mockup no. 2 at NAA during a critical design review of crew equipment stowage on February 9. At the time, Grissom is unofficially assigned as command pilot of Apollo 1, with Schweickart penciled in as backup pilot. Cunningham is in line for the Apollo 2 prime crew.

Left to right: Grissom, Schweickart, and Cunningham inside the mockup. The three astronauts also practice ingress and egress procedures. This mockup would be sent to KSC almost exactly one year later to aid in the Apollo I accident investigation.

Front to back: Astronauts Chaffee, White, and Jim McDivitt check out the mockup.

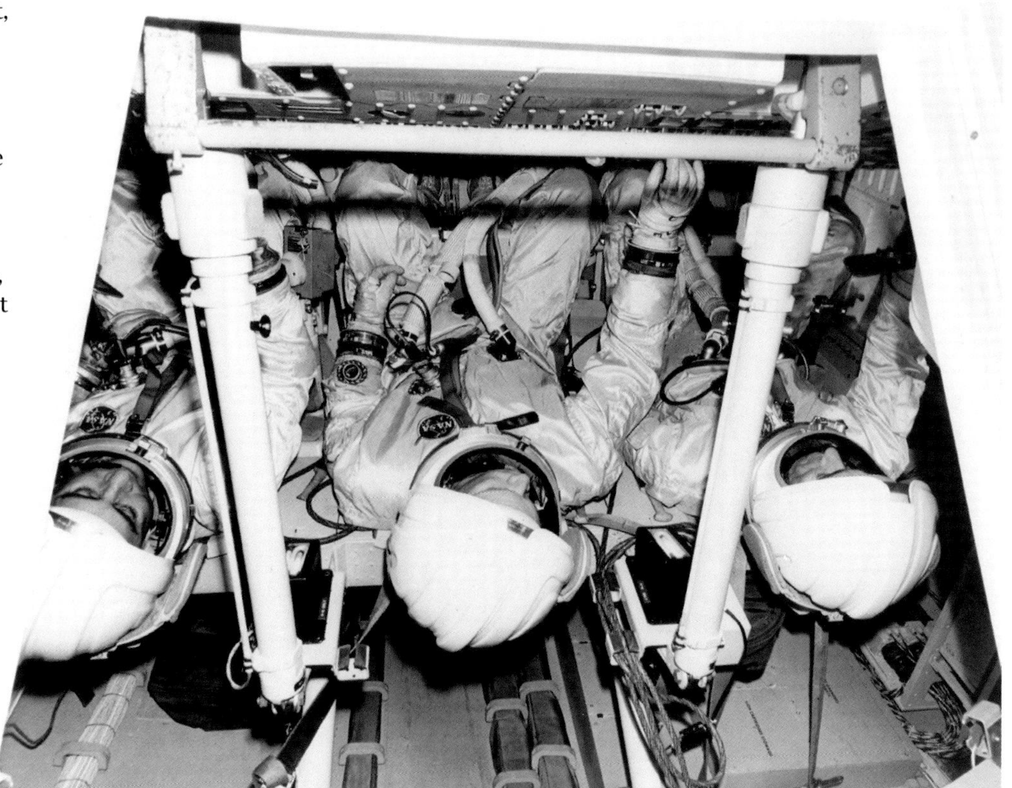

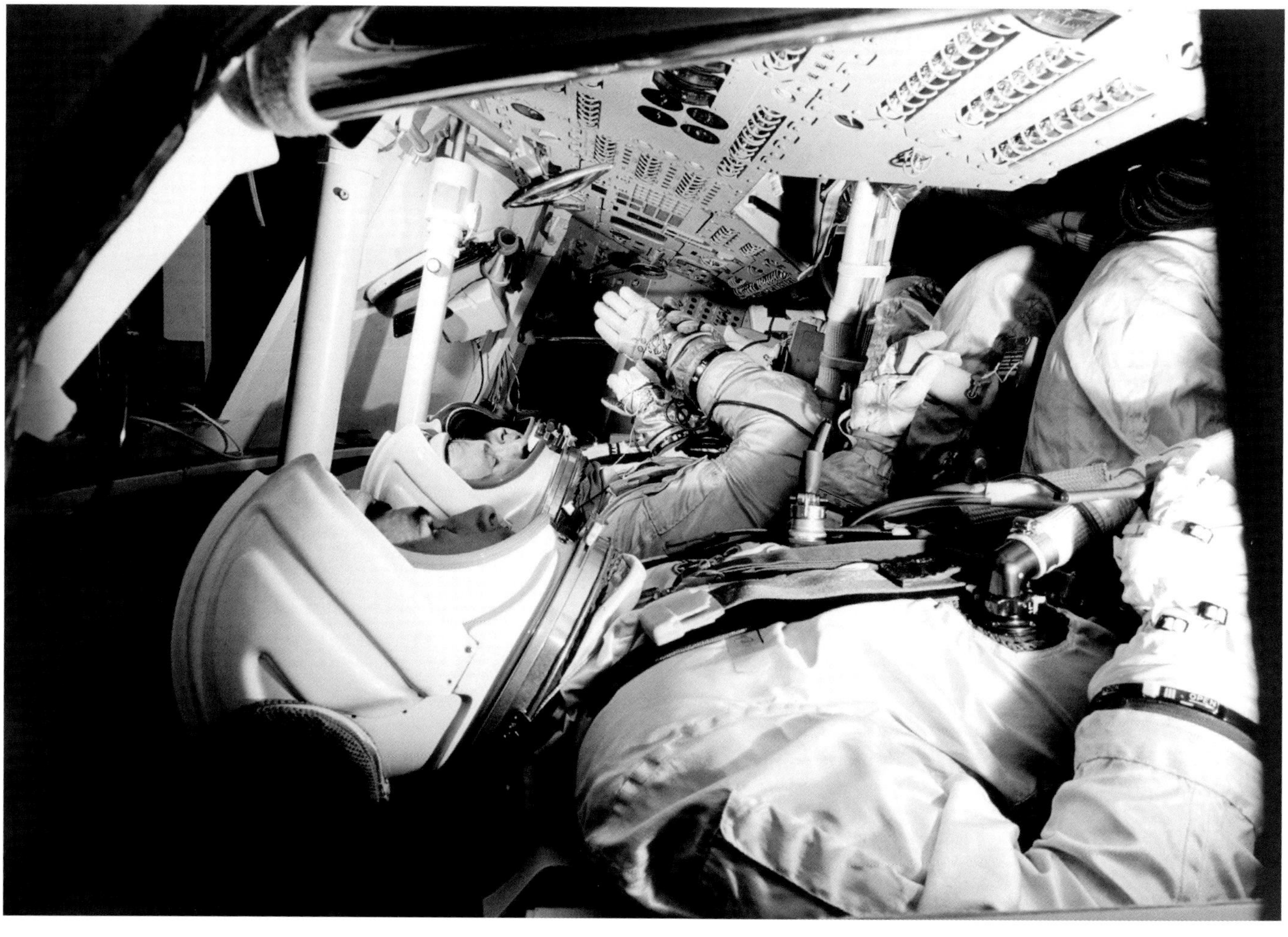

Left to right: McDivitt, Chaffee, and White pose on the stairs to the mockup. McDivitt is training as backup command pilot for Apollo I, while Chaffee and White are training as pilot and senior pilot, respectively, on the prime crew. Crew assignments, however, have not been officially announced.

The first Saturn IB (AS-201) sends unmanned Apollo CSM-009 on a thirty-seven-minute, 5,500-mile suborbital ride from Cape Kennedy AFS to the vicinity of Ascension Island in the South Atlantic. Launch is from LC-34 at 11:12 a.m. EST on February 26. Designed to determine if the new CM could survive the heat of atmospheric reentry, the test flight is nearly perfect.

The launch is photographed from three NASA T-38 jets flown by astronauts. Grissom and Chaffee photograph the booster above 30,000 feet, and McDivitt and White photograph the flight between 20,000 and 30,000 feet. Cunningham and Schweickart fly above the 37,000-foot level. Chaffee, White, and Cunningham pilot the two-seaters from the rear cockpit, while the others do the camera work from the front.

The CM is hoisted aboard USS *Boxer* after splashing down 35 miles from the aircraft carrier. The heat shield streaks have been caused by ablation of the filler compound covering bolts securing the shield to the CM. The circles mark the location of compression pads providing the seating between CM and SM. The box dangling from a cord (*at right*) contains the dye marker and had been ejected after water impact.

The Apollo I prime and backup astronauts take part in classroom instruction at MSC on March 1, three weeks before the crew is officially announced. *Left to right*: Grissom, Schweickart, McDivitt, training officer Raymond Zedekar, Chaffee, and White. At left is an Inertial Measurement Unit containing three gyros, used on both the CM and lunar module (LM).

White and Chaffee examine a mockup of the LM ascent engine in Building 6, the Gemini News Center, near MSC on March 21 before the announcement of the prime and backup crews for Apollo I. Behind them, McDivitt laughs with MSC director Robert Gilruth. Grissom (*behind White*) chats with MSC public affairs chief Paul Haney.

Backup command pilot Jim McDivitt deadpans an answer at the news conference. *Left to right*: Joe Shea, Apollo spacecraft manager; backup pilot Rusty Schweickart; McDivitt; prime crew members Chaffee, White, and Grissom; Gilruth; and Haney. Backup senior pilot Dave Scott, who had just wrapped up his Gemini VIII flight, does not attend.

Left to right: Chaffee, White, and Grissom react to reporters' questions. "I'm extremely pleased to be named," says Chaffee. "I think it will be a lot of fun."

Left to right: Chaffee, White, and Grissom. "I'm real pleased to be on the first flight; looking forward to it," says Grissom. "I've just now started getting into the Apollo systems, trying to forget everything I knew about Gemini."

Left to right: White, Grissom, and Chaffee after the news conference. NASA tells reporters a November two-week mission is possible, including perhaps a rendezvous with Gemini XII.

Chaffee receives an Air Medal and citation at MSC on March 25 for his efforts during the Cuban Missile Crisis. He was a photographic plane commander in Heavy Photographic Squadron 62 and completed eighty-two classified operational missions. *Left to right*: Chaffee's wife, Martha; Chaffee; Gilruth; Astronaut Office managers Alan Shepard and Deke Slayton; and deputy MSC director George Low.

NAA technicians prepare to use a scissor lift to move the service propulsion system (SPS) engine, an AJ10-137 manufactured by Aerojet General, toward the aft end of the SM for Apollo I on March 28. The engine's primary purpose would be to send the CSM into and out of lunar orbit. It was a simple engine that used two fuels that ignited when mixed.

In place, the engine is surrounded by the SM's plumbing, which will be protected by a heat shield. The SPS, which provides 20,500 pounds of thrust and could be gimbaled, allows course corrections during translunar missions and slows the spacecraft for reentry for earth-orbital flights.

NAA's Al Moyles (*left*) and Norman Abell prepare to enter a Block I CM in NAA's vacuum chamber in Downey on March 29 for the first extended test of Apollo's full life-support system. Richard Erman is already inside. They will spend two weeks on a simulated lunar mission.

Left to right: Erman (commander), Moyles (pilot and navigator), and Abell (systems engineer) get settled inside the CM before the simulation begins.

The test crew meets with the Apollo 1 astronauts at the end of their "mission" on April 12. *Left to right*: Moyles, Erman, astronaut Frank Borman, White, Chaffee, Grissom, and Abell. Abell tells reporters he was pleased with the results: "I think we could have stayed in there another two weeks if necessary—but none of us wanted to." Borman is the unofficial backup command pilot for Apollo 2 at the time and had spent almost fourteen days aboard Gemini VII as command pilot four months earlier.

The Apollo 1 SM undergoes checkout at NAA on April 15. Three of the four reaction control system (RCS) thruster quads are visible, with white protective covers over three heat radiators.

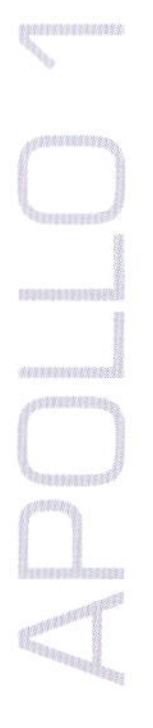

NAA technicians work on the inner pressure vessel of the Apollo I CM on April 18, which will be covered by the outer heat shields. One of five windows is at left, on the pilot's side of the spacecraft. Parachutes and inflatable landing bags will be packed around the cylinder at top, which will let astronauts transfer to their lunar module (LM) beginning with Apollo 9.

White holds a Saturn IB scale model (the Apollo I launch vehicle) during a visit to MSFC in Huntsville for three days of briefings on their booster. Lee James, Saturn IB program manager at MSFC, looks on at right with Grissom and Chaffee. Eight other astronauts also attend. Marshall is NASA's primary field center for rocket development.

Grissom and Chaffee go over the *Saturn IB Crew Familiarization* document at MSFC on April 19, 1966.

The crewmen attend a news conference at MSFC on the afternoon of April 19. *Left to right*: James, Slayton, White, Grissom, and Chaffee. The astronauts tell reporters that "complications on the Gemini VIII mission [ended prematurely by a thruster problem a month earlier] only proved that man can handle emergencies arising in space."

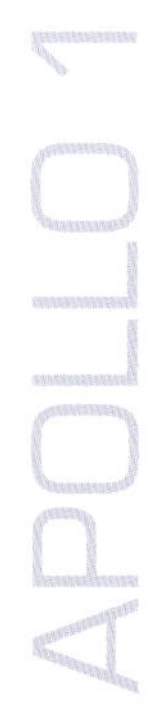

NAA workers at Downey prepare to hoist the Apollo 1 CM for a fit check on April 20. Side window no. 1 and rendezvous window no. 2 are left of the open hatch, with rendezvous window no. 4 at right.

The bridge crane moves the CM into position above the SM for a fit check without the aft heat shield attached.

Workers check the fit of the boost protective cover (BPC) apex to the forward heat shield on the Apollo 1 CM at NAA on April 27. The two "wells" in the cover are attach points for the launch escape system (LES) pylon.

The Apollo I astronauts (*top left*) pose with the NAA Apollo design team in front of a CM mockup at Downey in May.

Apollo I's Spacecraft / Launch Vehicle Adapter (SLA) is unloaded from the *Super Guppy* onto a scissor lift at the Cape Kennedy AFS Skid Strip on May 14. Its aft edge is protected by a light-blue ring. It has arrived from NAA's Tulsa, Oklahoma, facility.

Two cranes raise the SLA from the scissor lift as NAA workers steady it with taglines.

The SLA is lowered onto a transporter for the trip to the Manned Spacecraft Operations Building (MSOB) in the KSC Industrial Area on Merritt Island.

Hoses and cables surround the Apollo I CM at NAA on May 23. The light-blue structure provides support for installing and calibrating navigation equipment.

Workers operate with a constant reminder of the human life at stake in their products.

With the three couches not installed, Schweickart (*right*) is in the CM's lower equipment bay at NAA on May 23, where the navigation equipment is housed. The incomplete main display console is at top.

CHAPTER 3

June–July 1966

CM-007 was a Block I NAA production-line CM covered with cork on the aft and crew compartment heat shields to simulate the flight ablators. It was initially used in impact and acoustic vibration testing at Downey. In April 1966, it was the first CM delivered to MSC and was assigned to manned postlanding tests by the Landing and Recovery Division. The tests in the Gulf of Mexico included early operational and crew compatibility tests of the uprighting, environmental control, and communications systems.

The Apollo 1 crew undergoes recovery training aboard CM-007 in the Ellington AFB, Texas, swimming pool southeast of MSC on June 1. The CM floats with its apex down, known as the stable II position, with the three occupants suspended face down from their couches; three flotation bags would right the CM in practice to stable I if needed.

As divers supervise, Chaffee exits the CM in its upright (stable I) position, with White in a recovery raft on June 1. The astronauts practice egressing both in stable I and stable II.

White (*foreground*), Chaffee (*left*), and Grissom inflate their yellow water wings during the training, which would be repeated under more-realistic conditions in the Gulf of Mexico near Galveston, Texas.

White, with water wings inflated, floats in the pool.

The crewmen inspect the docking tunnel, which would be surrounded with parachutes and mortars. It could be used for crew egress if the side hatch could not be opened after splashdown, though not intended for Block I CMs.

Grissom (*center*) wears dark goggles to simulate a night recovery, with White at left and Chaffee in the background at right.

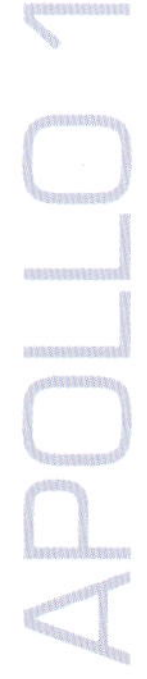

NAA developed two versions of the Apollo CM under a "building block" approach. *Block I* spacecraft were intended as test flights in Earth orbit, while the *Block II* versions would be for lunar missions. Two unmanned Block I CMs were launched by Saturn I boosters in 1966, and two planned manned Block I flights—Apollo 1 and Apollo 2—were canceled after the Apollo 1 fire. The Block II spacecraft then received significant safety changes—its hatch was redesigned, its cabin atmosphere at launch was changed to 60 percent oxygen and 40 percent nitrogen, and flammable material was removed or replaced. It also carried a docking mechanism for the LM starting with Apollo 9. The first manned Block II test would be Apollo 7.

NAA manager Norm Casson (*left*) chats with Grissom and White at Downey. Casson is responsible for testing and checkout of all CMs before they are sent to KSC, as well as for NAA's postrecovery operations.

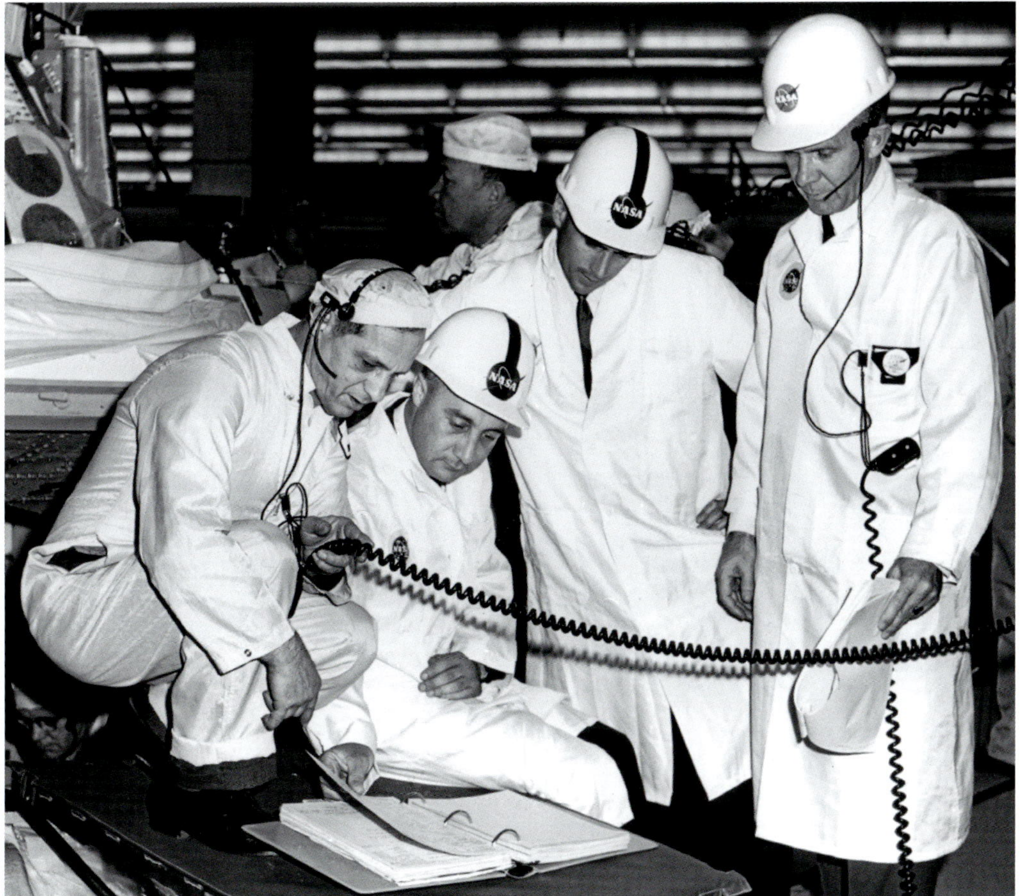

The astronauts visit NAA on June 23 and 24; the next twenty photos are from this trip. Grissom (*seated in center*) and Chaffee (*leaning over his shoulder*) chat with NAA engineers in Building 1 at Downey. At lower left, a technician seen through the hatch works inside the Apollo 1 CM.

Left to right: Grissom, Chaffee, and White examine the inner pressure shell of a CM, which will be covered by an outer heat shield. Openings in the shell at upper left are for the sextant and telescope.

Chaffee checks out the interior of the CM's crew compartment shell. Rectangular panels at upper left cover the sextant and telescope openings.

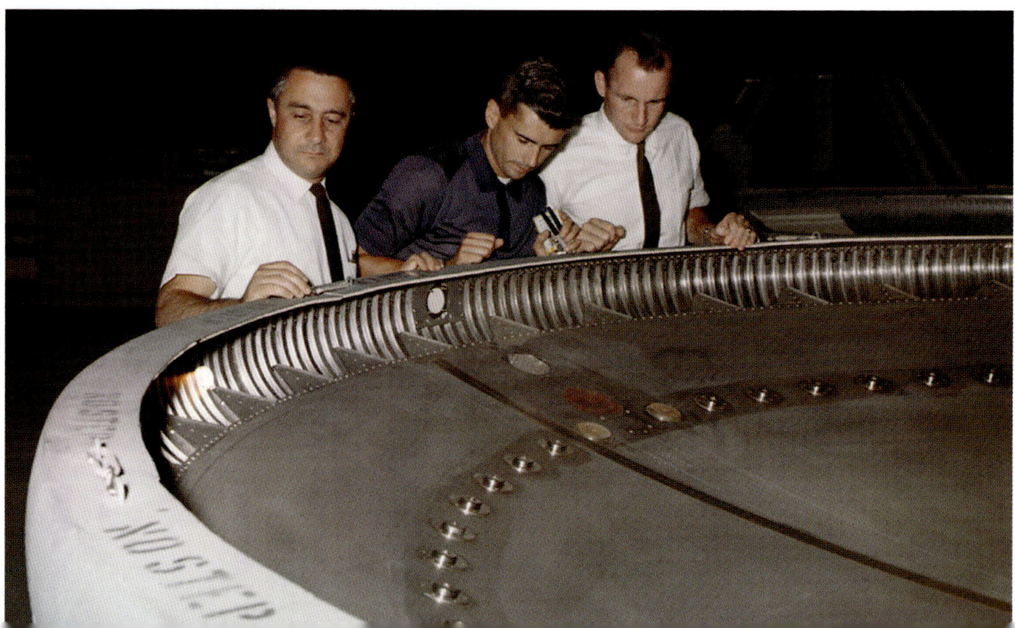

Left to right: Grissom, Chaffee, and White look over an aft heat shield. The bolts to secure it to the CM's aft bulkhead form an interior circular pattern. The frame for another CM is in the background at right. The CM is the first spacecraft designed to reenter Earth's atmosphere at lunar-return velocity, generating up to 5,000 degrees Fahrenheit.

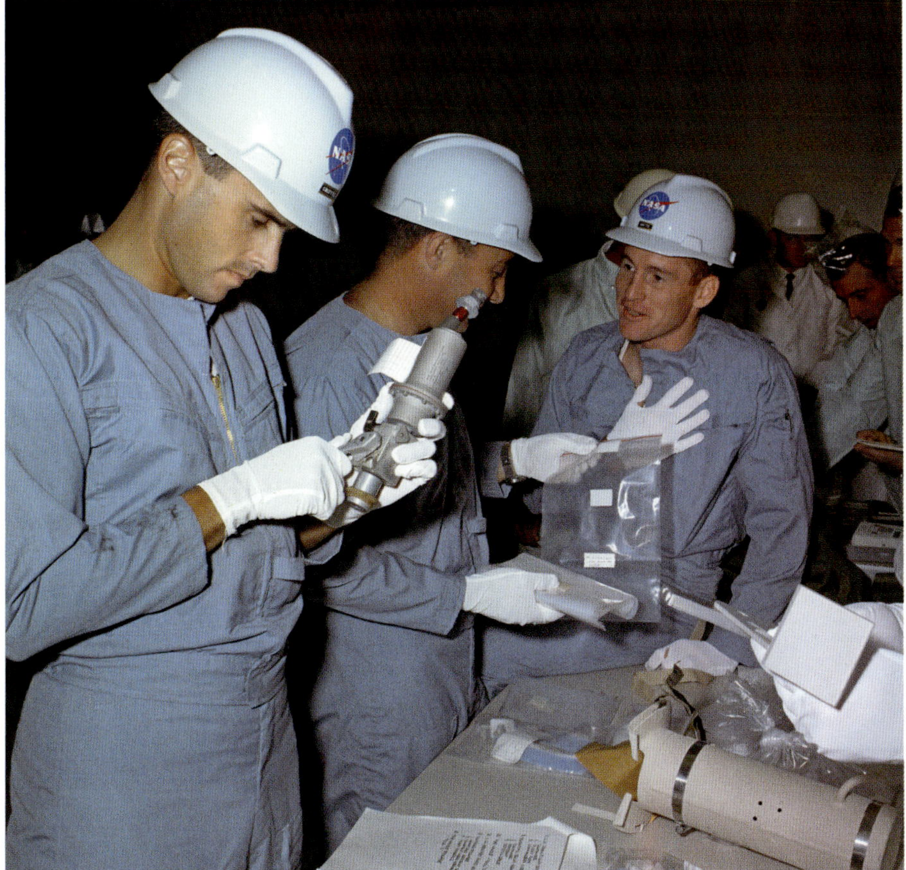

Left to right: White, Chaffee, Scott, Grissom, and Schweickart are briefed on pressure suit stowage by NASA suit technician Alan Rochford. Backup command pilot Jim McDivitt is not present.

Chaffee (*left*) holds a component from the CM's Environmental Control System with Grissom and White at right.

Left to right: Chaffee, Grissom, and White go over documentation. They wear prototype coveralls to be used during the flight after removing their space suits.

Grissom and White examine suit equipment with an A-1C spacesuit helmet at left.

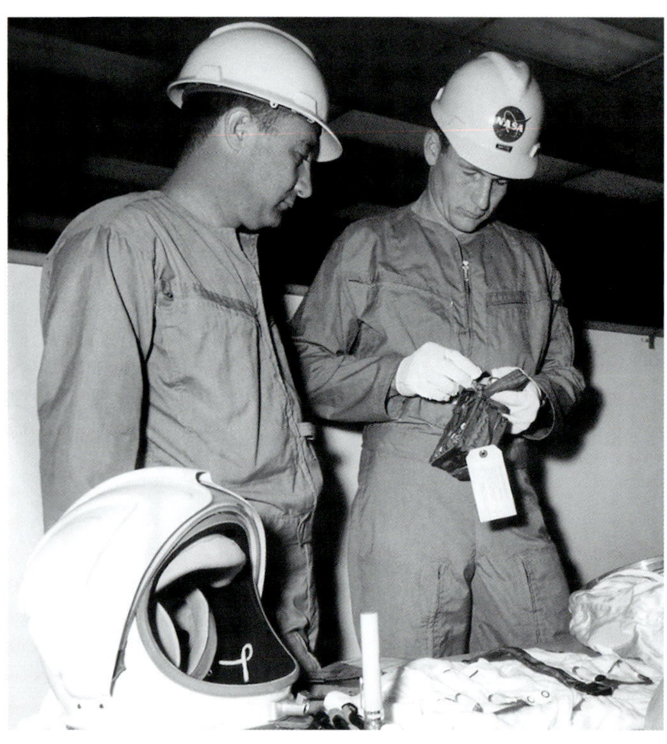

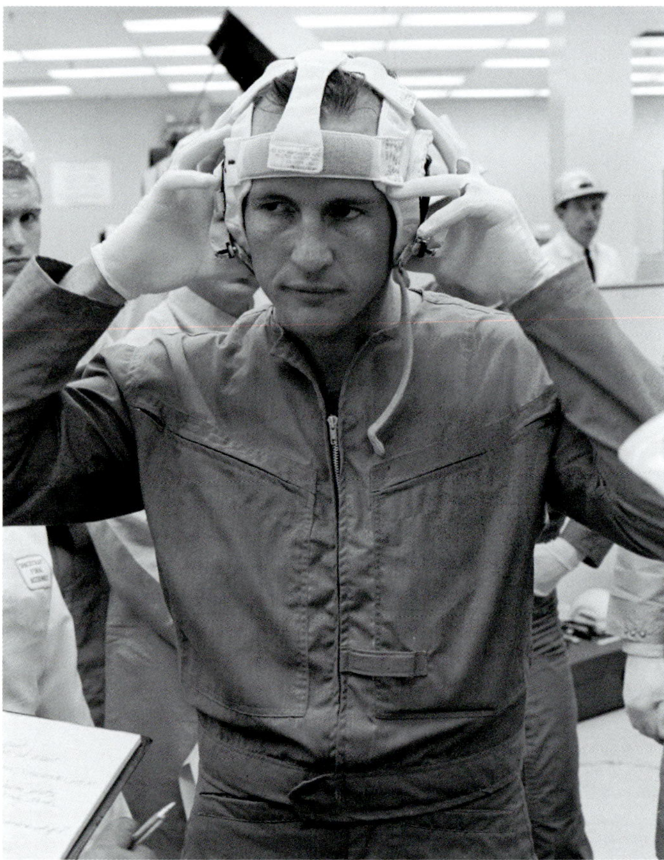

White tries on a prototype communications cap during the suit equipment inspection.

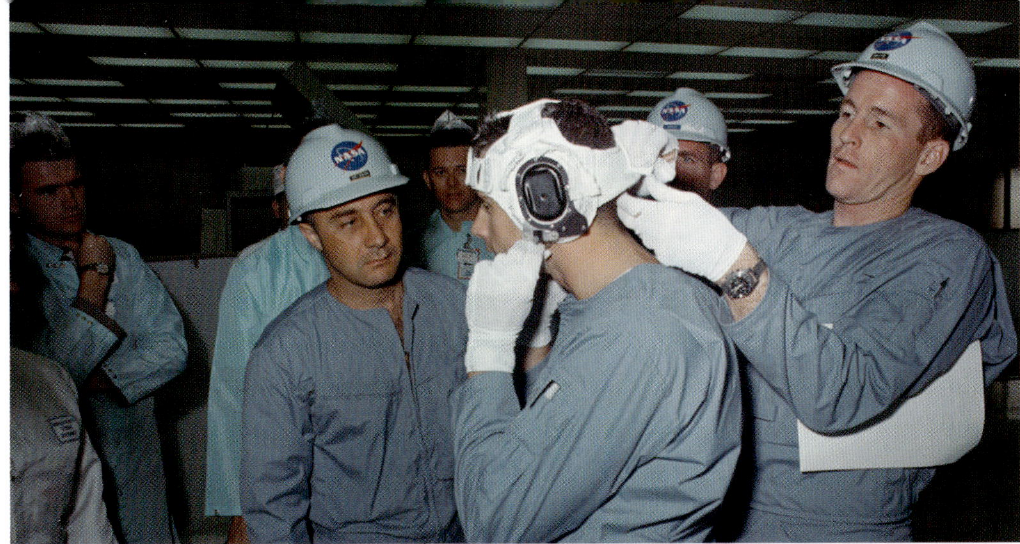

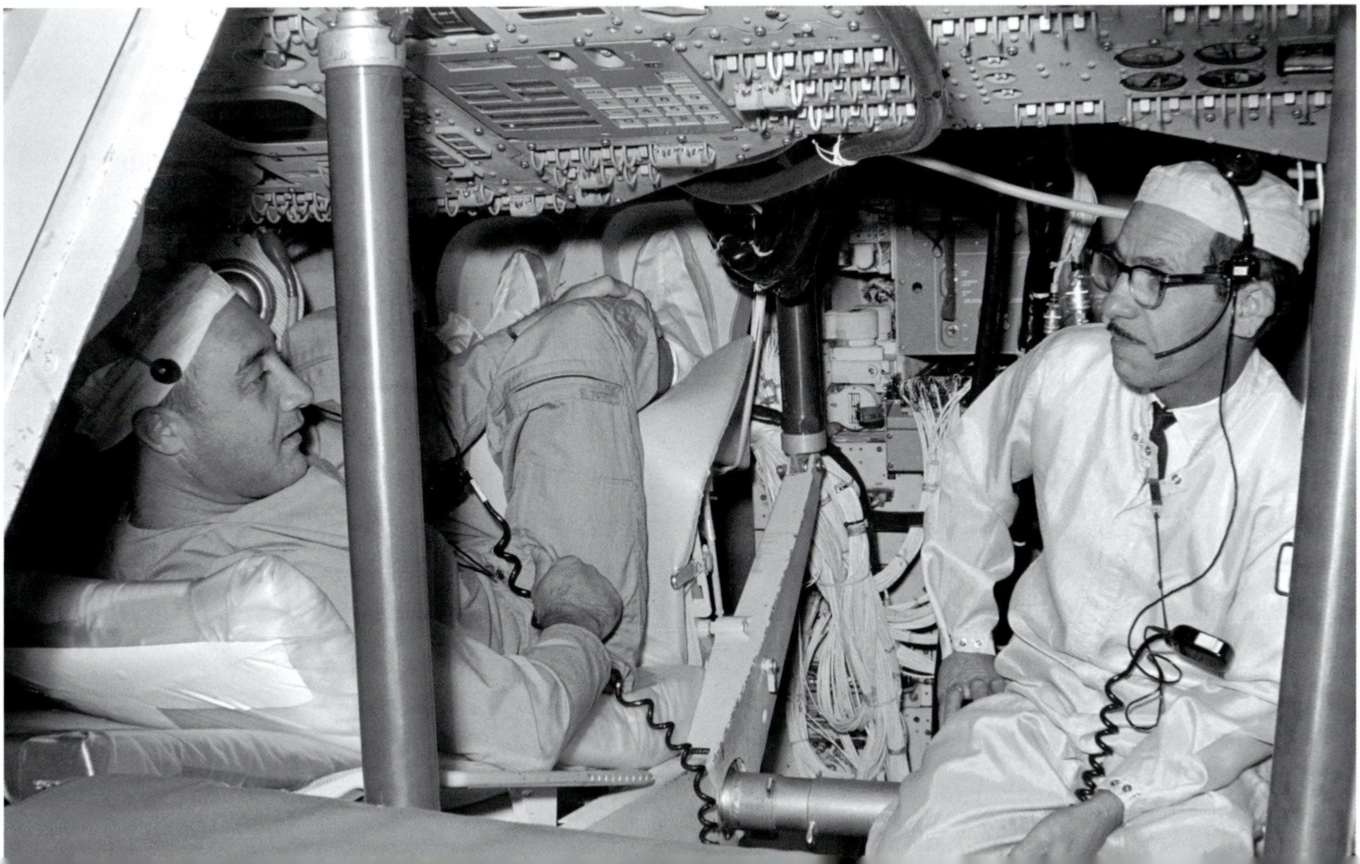

White assists Chaffee with the headgear. Looking on are Rochford (*left*), Grissom, and Scott (*partially obscured behind Chaffee*).

White (*left*) and Grissom inspect a CM under construction with an NAA engineer.

Grissom (*left*) in the command pilot's left-hand couch with an NAA technician. The senior pilot would occupy the center couch (removed here) with the pilot on the right.

Grissom and White (*center*) chat with an NAA quality control inspector outside Apollo I. Two technicians visible through the open hatch work inside.

Grissom lies across the hatch sill, with an NAA inspector inside.

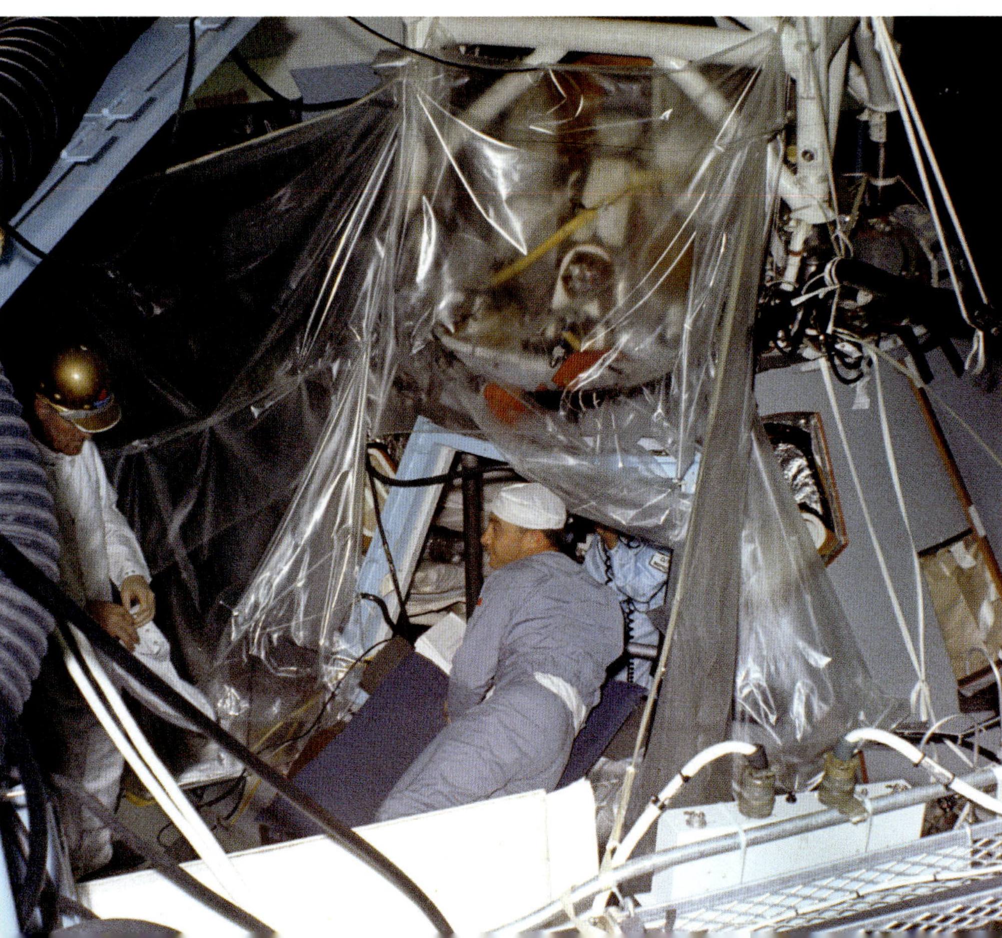

Left to right: Chaffee, Grissom, and White pose with a CM mockup.

Left to right: Chaffee, Grissom, and White pose beneath a mockup of the SM and its SPS engine nozzle.

Front to back: Grissom, White, and Chaffee are seen through a side window of the mockup.

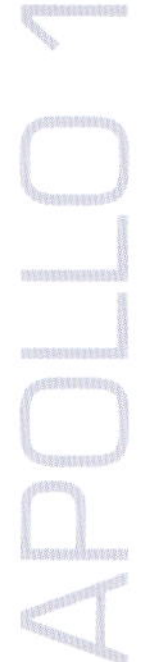

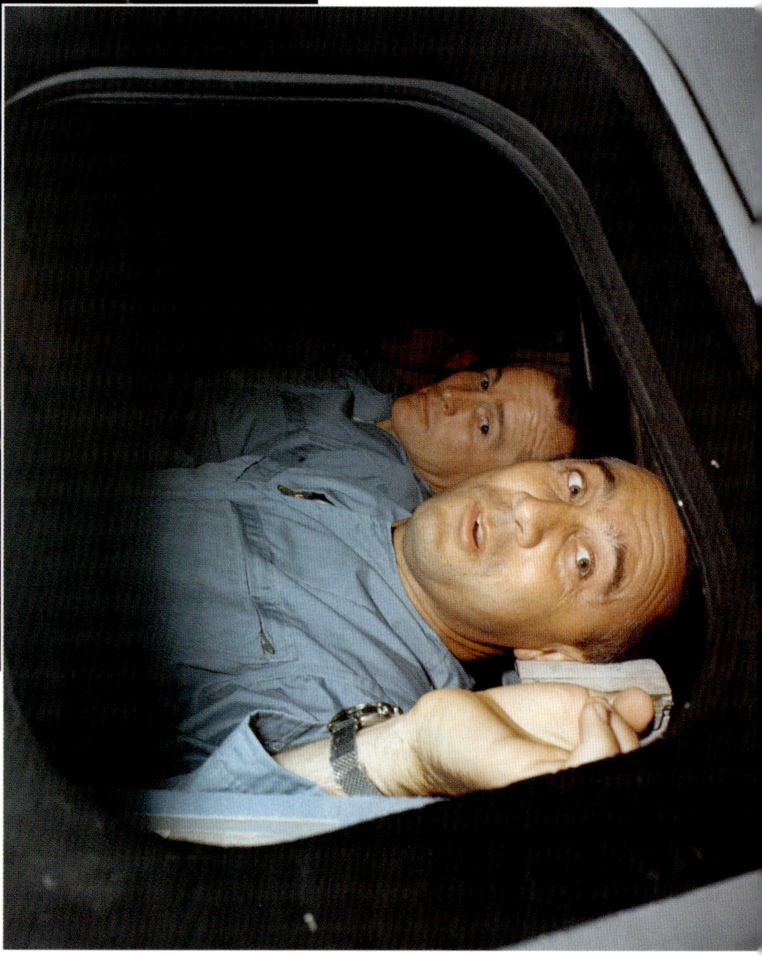

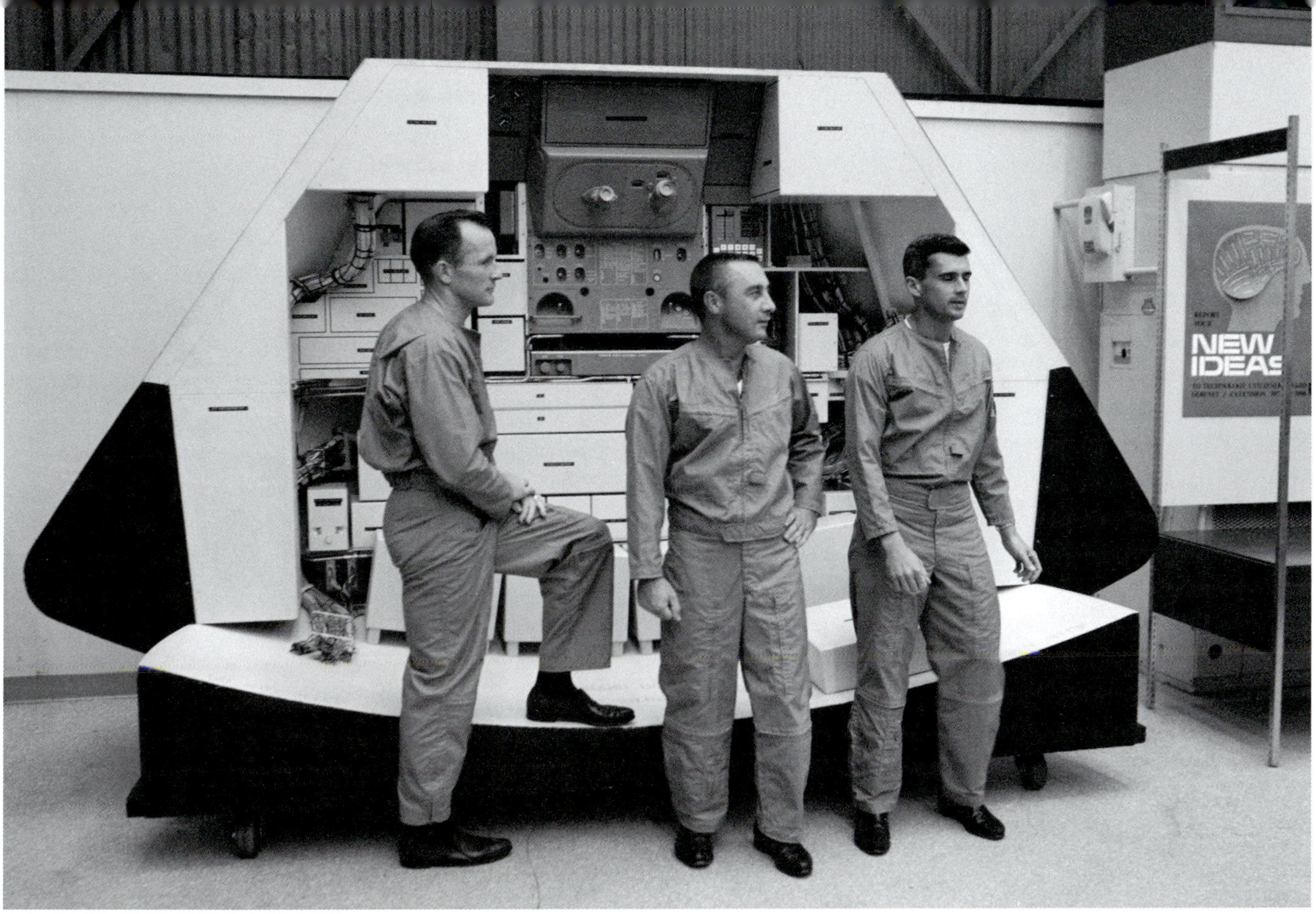

Left to right: White, Grissom, and Chaffee pause in front of a CM cutaway. The lower equipment bay (*behind Grissom at top*) includes the sextant, telescope, and other navigation equipment.

Workstand scaffolding surrounds the Apollo I CSM in Building 290 at NAA during integrated systems testing by the Apollo Test Office. The easels at left show test team technicians at work, with crew portraits at right.

White (*left*) and Grissom outside the workstand surrounding the Apollo I CSM at NAA. "Hazardous testing" reads the sign between them. "Your presence is subject to challenge."

The countdown for the second Saturn IB (AS-203) is underway at LC-34 on the evening of July 4 for launch the next morning. The goal is to increase confidence that its liquid hydrogen–fueled S-IVB second stage could be restarted in the microgravity of space, a requirement to push manned Apollo spacecraft out of Earth orbit toward the Moon.

AS-203 launches at 10:53 a.m. EDT on July 5. The S-IVB and its Instrument Unit and nose cone will separate from the first stage as one body, 92 feet long and weighing 58,500 pounds—the heaviest "payload" ever orbited. After providing data about how the liquid hydrogen fuel behaved during four orbits, the S-IVB is intentionally exploded to better understand the pressure limits of its fuel tanks.

John Hodge, NASA's second flight director, monitors the AS-203 mission from his console at the Mission Operations Control Room (MOCR) at MSC. A British aerospace engineer, he joined NASA's Space Task Group in 1959 after a jet interceptor project in Canada was canceled.

A bridge crane prepares to raise Apollo 1's SLA no. 5 after it arrives in the MSOB on July 7. The cylindrical structure at left is altitude chamber L.

The 28-foot-tall SLA is moved to Integrated Test Stand no. 1 at left. The blue ring protects the bottom surface until mating at the pad. The adapter would house an LM during ascent for Apollos 9–17 but serves as a structural interstage between the SM and the IU for Apollos 1, 7, and 8.

The SLA is positioned in the test stand. The 3,800-pound structure is made of aluminum honeycomb material covered by a thin layer of cork painted white to minimize thermal stresses during launch and ascent.

CHAPTER 4

Launch Complex 34

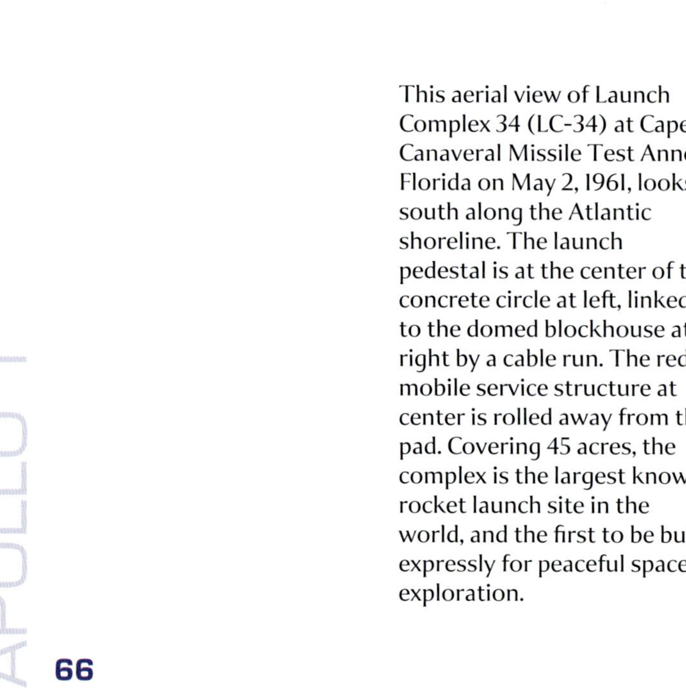

This aerial view of Launch Complex 34 (LC-34) at Cape Canaveral Missile Test Annex, Florida on May 2, 1961, looks south along the Atlantic shoreline. The launch pedestal is at the center of the concrete circle at left, linked to the domed blockhouse at right by a cable run. The red mobile service structure at center is rolled away from the pad. Covering 45 acres, the complex is the largest known rocket launch site in the world, and the first to be built expressly for peaceful space exploration.

This view looks southeast on July 11, 1961. Just south of the blockhouse, the Operations Support Building is under construction, which would house calibration and testing labs and offices. The liquid oxygen (LOX) storage tank is at left. In the distance at top left is the Cape Industrial Area. The complex was built under the supervision of the US Army Corps of Engineers at a cost of about $6.2 million, beginning in June 1960, with acceptance by NASA on January 10, 1961.

The first rocket of the Saturn family, SA-1, is on the pad in September 1961 surrounded by its service structure, with the two-story blockhouse in the foreground. This Saturn I booster with two dummy upper stages would launch on October 27, a suborbital flight that is the first of the Apollo program. A separate umbilical tower would be added to the pad in 1962.

The interior of the second floor of the blockhouse nears completion in March 1961. A periscope allows a safe view. The first floor includes tracking, telemetry, and communications consoles for booster and upper stages. The facility, 1,000 feet from the pad, is officially known as the Launch Control Center. The interior dome was sprayed with a 2-inch coat of acoustical material, and the exterior is covered with reinforced concrete 5 feet thick, topped by a thick layer of soil and more concrete.

Launch Operations Directorate chief Kurt Debus (*lower right*) monitors the countdown for SA-3 on November 16, 1962, at the test supervisor's console in the blockhouse. It is the third of four successful Block I Saturn I launches.

The service structure surrounds the final Block I Saturn I to launch from LC-34 on March 18, 1963, which would lift off ten days later. The structure provides work platforms and two traveling cranes. At 2,800 tons and 310 feet tall, it is the largest movable land structure in North America and houses four elevators. It moves on four twelve-wheel "trucks" along a special dual-track railway within the complex. The next Saturn I test launches use the Block II configuration, with longer fuel tanks and improved H-1 engines to increase thrust.

Ten Saturn I rockets (four Block I and six Block II) were launched from 1961 to 1965 before the booster was replaced in 1966 by the heavy-lift derivative, the Saturn IB (NASA briefly designated it as the Uprated Saturn I). It used a slightly larger and more powerful version of the Saturn I first stage, and the Saturn I's S-IV second stage was replaced by the S-IVB. It used a single Rocketdyne J-2 engine instead of six RL-10 engines and had an improved guidance and control system. The S-IVB would also be used as the third stage of the Saturn V Moon rockets.

New consoles are ready for installation in September 1963 to update the blockhouse for Saturn IB launches, which would begin in 1966. A second periscope has been installed. The IB had been added to the program in 1962 to launch early manned Apollo missions into Earth orbit without the weight of an LM.

The launch umbilical tower, added during the summer of 1962, rises 240 feet from the pad, which is 430 feet in diameter, in this November 20, 1963, photo. The tower's three swing arms support electrical cables and hydraulic and pneumatic lines to the booster. Twin flame deflectors are parked at the edge of the pad at right. Part of the pad is covered with refractory brick to minimize heat damage from launch.

A 1964 Ford Mustang pauses on October 18, 1964, near the service structure at its parked position, with the umbilical tower in the background. During Sunday afternoon tours permitted by the USAF starting in December 1963, the public could drive private vehicles on a predetermined route that provides views of launch pads and facilities.

A sign along the tour route identifies the turn from ICBM Road to LC-34.

Wackenhut "guardette" Nancy Conhen poses next to the tourist information trailer on December 1, 1964. "When I tell [visitors] we are planning escorted bus tours in about 18 months," she tells *Spaceport News*, "many of them express an interest in returning." This first facility is just west of the Gate 3 entrance to the Kennedy Space Center (KSC) on Merritt Island. TWA had recently begun a contract with NASA to operate the bus tours, as well as security and fire control through subcontractor Wackenhut.

The LC-34 service structure, shown in January 1965, received a major overhaul starting in October 1963 to prepare for Saturn IB launches. Four pairs of retractable hurricane doors and four weathertight silo enclosures were added. Eight vertically adjustable service platforms and a traveling hoist for the LES were also installed, and new anchor points were built to keep the then-heavier service structure in place over the pad during high winds.

Other changes included modifying the umbilical tower's existing three swing arms to meet the S-IB's larger dimensions. Also added were a fourth swing arm for crew access for manned missions, and a high-speed elevator to provide emergency crew egress. Total cost of the construction was $5.3 million.

The service structure is nearly touching the umbilical tower in this view looking southeast during construction work, which includes upgrades to LC-34's RP-1 and LOX fueling facilities. The blockhouse for LC-37 is at right in the distance. Cape Road leads to the Industrial Area in the distance at left.

A new umbilical tower swing arm (no. 4) at the 224-foot level will let astronauts cross from left to right to enter a white room to provide a small enclosure for the crewmen and the technicians helping them ingress the CM. While the next launch, AS-201, would be unmanned, one goal of the upgrades is to man-rate the complex.

This view of swing arm no. 4 from the opposite side shows a rotary hydraulic actuator (ROTAC), which would rotate each arm away from the vehicle at liftoff to its retracted position.

Claude Moses, KSC operations section chief, oversees LC-34 and six other launch complexes in 1966. The thirty-one-year-old engineer had worked on ballistic missiles in the US Army and joined NASA in 1959.

The twin flame deflectors show scorching on June 6, 1969. Each deflector weighs 150 tons and is 21 feet high. Their 1-inch-thick steel skin is covered by a special heat-protective ceramic coating. *Photo by Jacques Tiziou*

Leavenworth's tickseed, the Florida state wildflower, blooms along the cable run on June 6. LC-34's final launch, Apollo 7, had come in October 1968. The service structure and umbilical tower were demolished in 1972. *Photo by Jacques Tiziou*

The LC-34 blockhouse was a stop on the Cape Canaveral tour, as seen in January 1980, until the mid-1980s. The facility was later abandoned. *Photo by J. L. Pickering*

A sign chronicles the complex's seven launches, photographed in June 1979. *Photo by J. L. Pickering*

The blockhouse interior, photographed in June 1979. *Photo by J. L. Pickering*

CHAPTER 5

August–September 1966

Chrysler Space Division technicians inspect both ends of the 82-foot-long first stage (S-IB-4) of Saturn IB AS-204 at NASA's Michoud Assembly Facility in New Orleans, Louisiana, in early August. Eight Rocketdyne H-1 engines surrounded by flame curtains with their protective throat seals installed are mounted to its tail section. The stage would soon be barged to Cape Kennedy AFS for what would be called Apollo I.

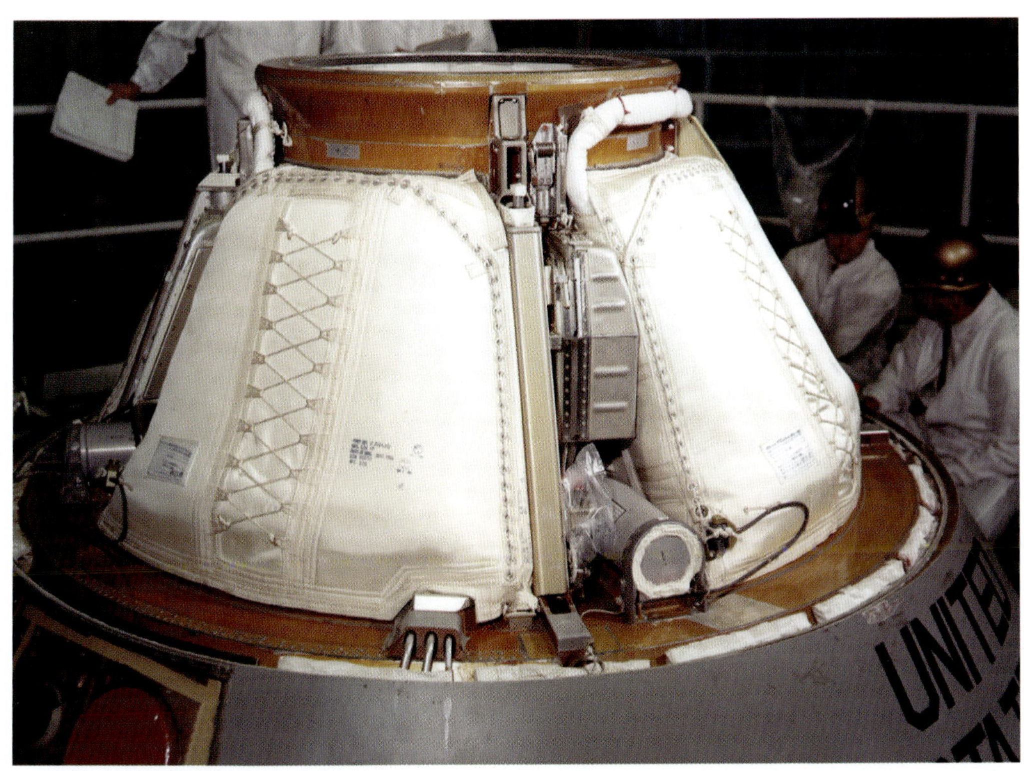

Two of the three vacuum-packed main parachute deployment bags surround the forward access tunnel of CM-012 at NAA in Downey, California, on August 1. The uprighting bags are stowed behind them. Two of three cylindrical pilot parachute mortars that will deploy the main parachutes protrude between the bags. The components are part of what is collectively known as the Earth Landing System.

NAA technicians prepare to raise the apex cover of the forward heat shield to the top of the integrated test stand for installation on CM-012 on August 1. The square opening is for a LES well; the second opening is to accommodate RCS pitch engines.

The Apollo I prime and original backup crews participate in a news conference at NAA in Downey on August 4. *Left to right*: Rusty Schweickart, Dave Scott, Jim McDivitt, Grissom, White, and Chaffee. At far left is MSC public affairs chief Paul Haney. A model of the Apollo CSM and LES is at left; at right is a model of the Saturn V launch vehicle and its mobile launcher.

Scott speaks during the news conference, with a full-scale CM mockup at right. “The crew has to log some 6,000 pieces of data during the flight,” he tells reporters. “I think there’s a good opportunity we’ll get it off this year,” Grissom says, adding, “It is still scheduled for next year, but we’d be foolish not to go before the first of the year if it’s ready.” The flight could last longer than fourteen days, says Chaffee, “if everything is going all right.”

White answers a question from ABC News science editor Jules Bergman (*right*) with Chaffee (*left*) and Grissom with a CM mockup on August 4 at NAA.

NAA technicians check the fit of the legs of the titanium tower of the LES into wells in the forward structure of CM-012 at Downey in August. Each leg is attached by a stud on a frangible nut, which is broken apart by two small detonators to separate the tower about two and a half minutes after launch. The test stand in the background surrounds CM-014, which is planned for Apollo 2 at the time. It would be partially disassembled for study at KSC after the Apollo 1 fire.

Apollo I's S-IVB (S-IVB-204) is moved from the Super Guppy onto a scissor-lift transporter at the Cape Canaveral AFS Skid Strip on August 7. It has arrived from the contractor Douglas Aircraft Corp.'s test facility near Sacramento, California. The Super Guppy had joined the smaller Pregnant Guppy in service the previous year.

The stage emerges from the aircraft. The four cannisters with hoses are part of the environmental control system during air transportation.

Douglas workers prepare to move the S-IVB, with its single J-2 engine covered by a tarp, to the Vehicle Assembly Building (VAB) at KSC. The KSC Transportation Branch, coordinating the move, is responsible for operating everything from custom equipment trailers, barges, and administrative aircraft to a taxi fleet and shuttle buses. The branch is staffed by contractor Bendix Corp.'s Launch Support Division.

CSM-008 is installed in the 65-by-120-foot chamber A of the Space Environment Simulation Laboratory at MSC in Houston in June. The spacecraft would undergo unmanned and manned altitude and temperature tests in the chamber.

Below: the three engineers who lived aboard CSM-008 for one week (*left to right*), Joel Rosenweig, Neil Anderson, and Donald Garrett, receive certificates at the conclusion of the test on August 9. They were assigned to the Flight Crew Support Division at MSC.

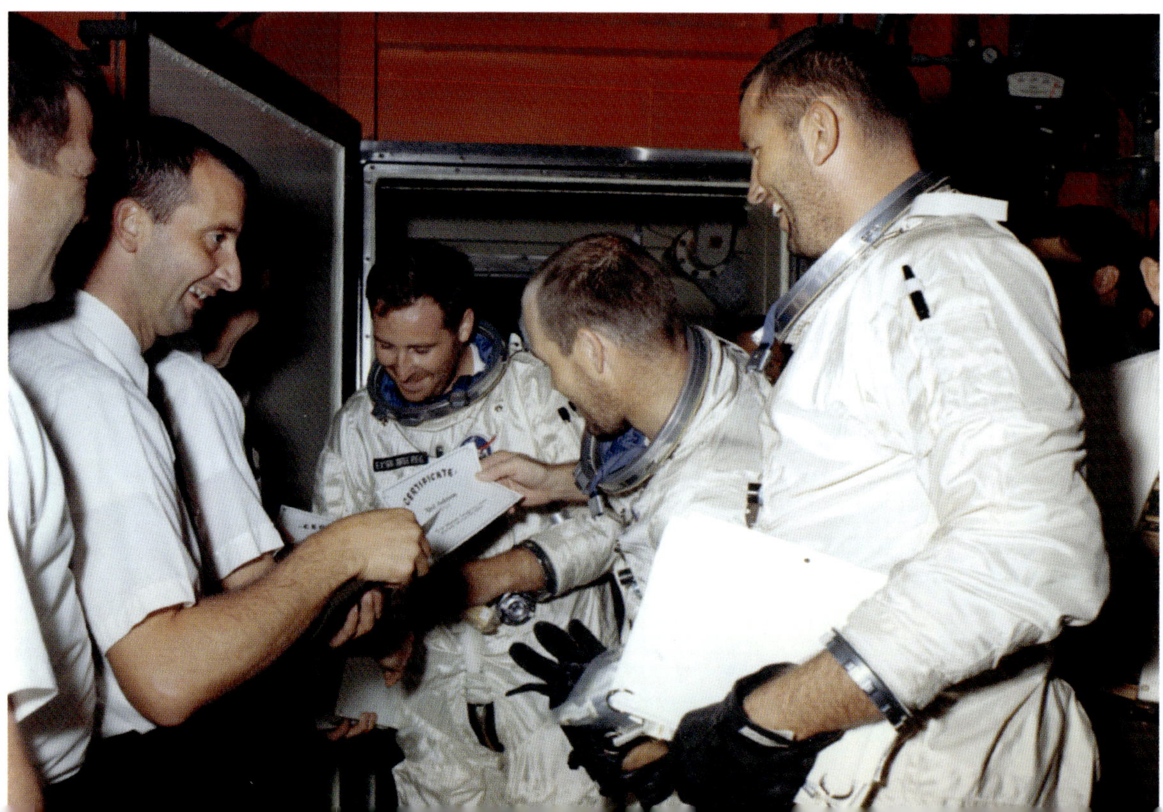

Workers secure protective covers over SM-012 at NAA on August 8 for shipment to Cape Kennedy AFS. Its large SPS nozzle extension would be bolted on after the SM passed an altitude chamber test at KSC.

The S-IVB-204 aft interstage arrives at Port Canaveral, south of Cape Kennedy, from NAA's Seal Beach, California, plant on August 9. The SS *Steel Executive* also carries components for the second and third stages of the first Saturn V. The twelve-day voyage was through the Panama Canal.

SM-012 is loaded aboard the Pregnant Guppy aircraft near Downey on August 9 for shipment to Cape Kennedy AFS.

SM-012 waits to be unloaded from the Guppy on its transporter, with a crate and boxes of associated hardware, at the Cape Kennedy Skid Strip on August 10.

SM-012, parked between sections of the Guppy, is ready for transport to the Cape Industrial Area.

SM-012 is moved into Hypergol Support Building in the Industrial Area for inspection and processing on August 10. The facility handles hypergolic fuel system components that don't pose toxic or explosive hazards.

The SM is ready to depart the Hypergol Support Building on a ground transporter on August 12.

SM-012 arrives at the MSOB in the KSC Industrial Area on Merritt Island for a series of tests.

The SM, with its radiators covered, is suspended from a bridge crane in the MSOB on August 12 before SPS leakage and function tests in the coming week.

On August 15, the S-IB first stage for Apollo 1 is towed from the NASA barge *Palaemon* at the Cape Kennedy AFS wharf. It had left the Michoud Assembly Facility the previous week.

The stage emerges from *Palaemon* on its transporter as a Technicolor cameraman films at left.

The S-IB is backed into Hanger AF, about 400 yards away, for inspection.

A crane lifts the Instrument Unit (IU) for AS-204 (S-IU-204) from its Super Guppy shipping cradle at the Cape Kennedy Skid Strip on August 16 after arriving from MSFC in Huntsville. Manufactured by IBM, the IU is the nerve center of the Saturn IB, containing guidance, tracking, and telemetry equipment.

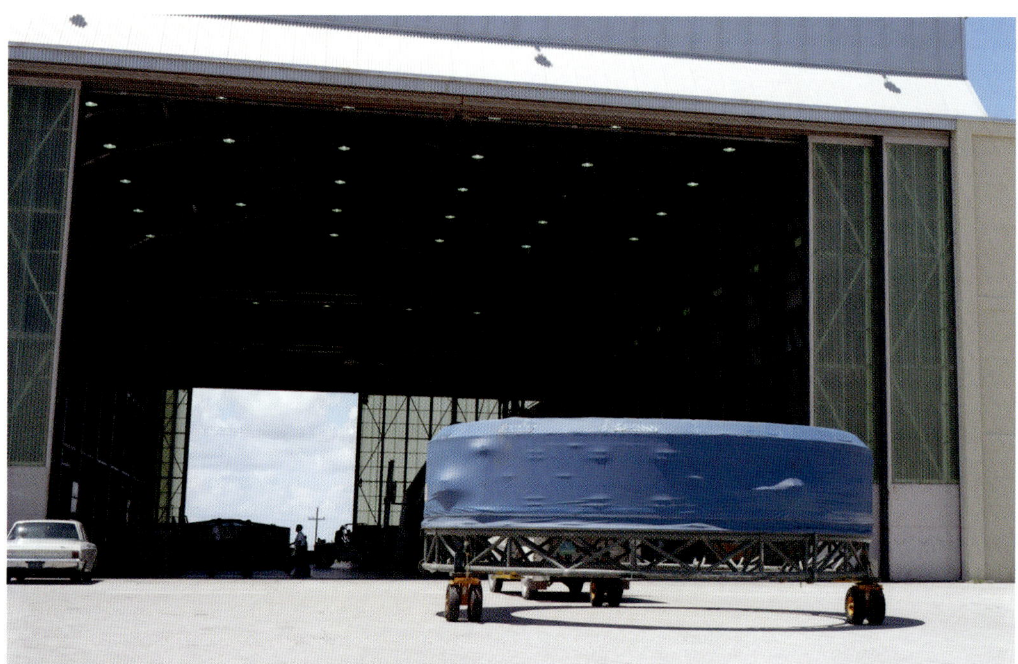

The IU is towed toward Hangar AF.

The IU is parked next to the first stage for AS-204, which had arrived the day before.

Bottom to top: Chaffee, White, and Grissom pose before entering their CM at NAA in Downey on August 18 for a final checkout before shipment to the Cape. Since no EVAs are planned, the astronauts wear a modified Gemini G3C suit, the AIC. This version adds new electrical and environmental disconnects and a protective shell for the Plexiglas helmet visor when open. It was the final suit made by the David Clark Co. of Worcester, Massachusetts, since ILC (International Latex Corp.) of Dover, Delaware, would get the contract for the rest of the Apollo program. The panel at left monitors the spacecraft's environment.

Left to right: Grissom, Chaffee, and White prepare to enter CM-012 at Downey on August 18.

White waits to ingress the spacecraft next to Norm Casson, the NAA manager responsible for CM testing and checkout.

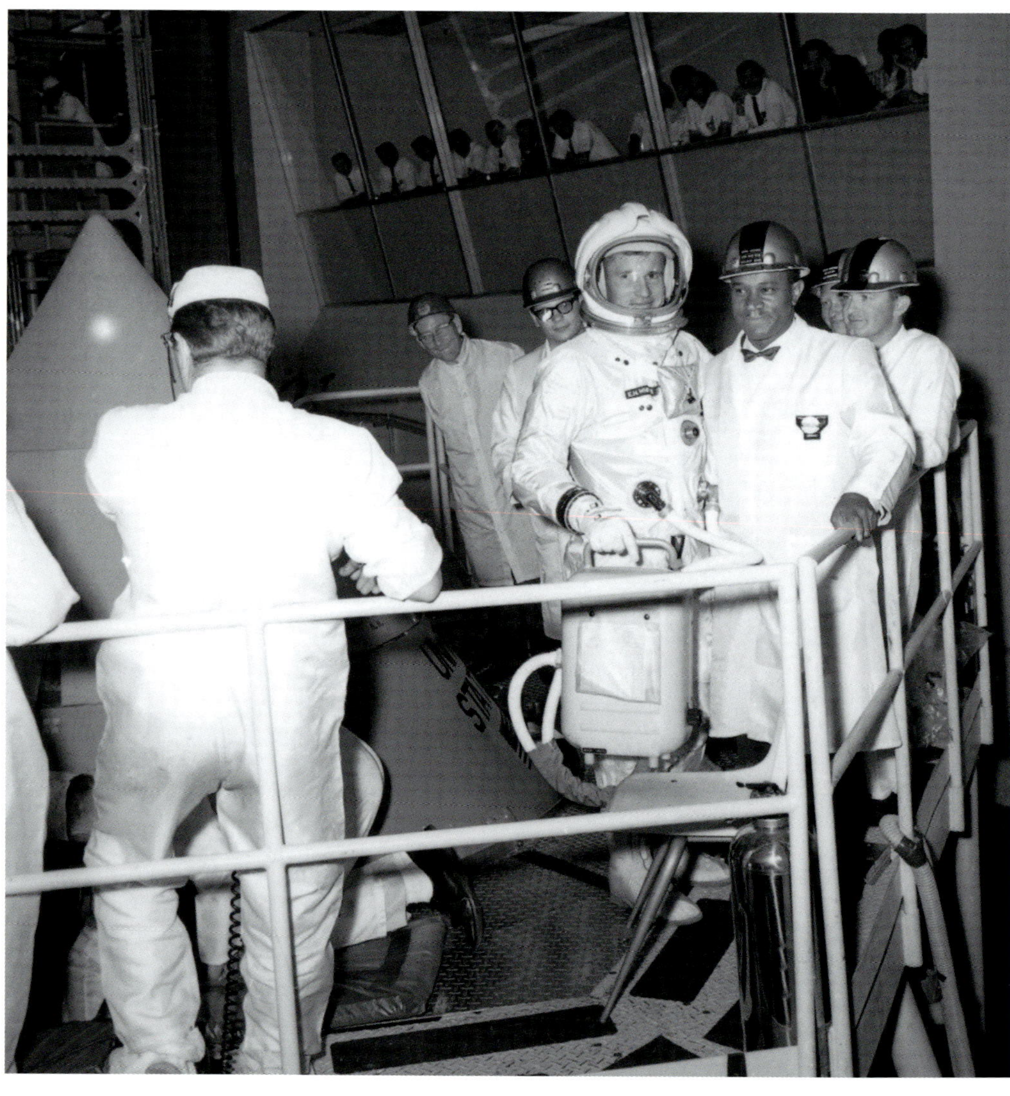

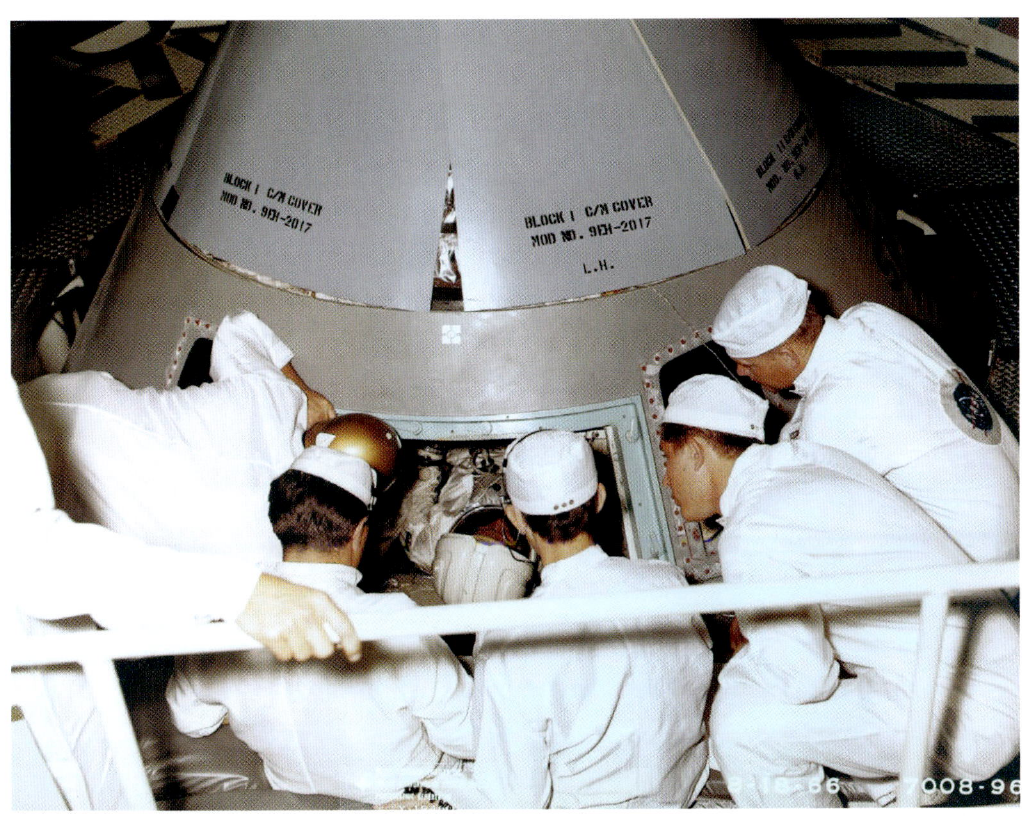

NAA and NASA technicians surround the hatch, with White in the center couch.

NAA and NASA personnel in clean-room garb pose with CM-012 in the high bay area of Building 6 at NAA on August 24, just prior to shipment to Cape Kennedy AFS. Quality inspectors wear green smocks.

NAA personnel prepare to pack up CM-012 on August 24. The curved red cover at right protects a scimitar radio antenna. The exterior is left a medium flat gray, the color of the epoxy resin heat shields covering the CM. NASA accepts the spacecraft despite 113 engineering changes still pending.

Workers load the CM aboard the Pregnant Guppy aircraft near Downey on August 24 for the trip to Cape Kennedy. The wheeled transport could raise the platform holding the spacecraft and its portable environmental control unit so the CM could be moved into the plane on rails.

CM-012 shortly after being unloaded at the Skid Strip at Cape Kennedy AFS on August 25. The service structures for two pads along ICBM Row can be seen on the horizon. NASA officials are not pleased that NAA has labeled the CM "Apollo One," since the decision to include astronauts has not been formally announced nor the name approved. The placard has been signed by Downey workers.

The CM and a crate containing associated hardware are lowered for the trip to Hangar AF. LC-34 is in the distance at center.

AS-202's Saturn IB heads skyward from LC-34 at 1:16 p.m. EDT on August 25. This third and final developmental Saturn IB launch carries CSM-011 on a successful suborbital flight. Primary objectives include demonstrating CM and SM separation and testing the heat shield at high reentry speeds simulating a lunar return. The Apollo 1 crewmen are spectators.

Workers ride behind the Apollo I CM on the rear of the trailer as it starts to head toward Hangar AF.

The CM arrives at NASA's Hangar AF in the Cape Industrial Area. Originally known as the Special Assembly Building, it had provided checkout facilities for Gemini and also houses Saturn administrative support offices.

Workers begin to uncrate the CM inside the hangar the same day. It would be sent to the Pyrotechnic Installation Building (PIB) the following day for weight and balance testing, then moved to the MSOB on August 29. After the fire, it would be disassembled in the PIB.

AS-204's S-IB first stage stops at the LC-34 perimeter gate on August 29. The four outboard engines are mounted on gimbals, allowing them to be steered to control the booster's flight. The blockhouse is at center.

Chrysler workers remove the tarp and attach pad cranes to the stage.

Workers use cranes and taglines to tilt the stage. Only three of its eight fins, which provide launch stand support, hold-down attach points, and aerodynamic stability in flight, were installed for transportation. The stage's heat shield surrounds the eight Rocketdyne H-1 engines, with their red throat seals still in place. The launch complex's two flame deflectors are parked in the background at center.

The S-IB is ready to be raised vertically. The stage is composed of eight Redstone tanks alternately painted black (RP-1 fuel) and white (LOX) for thermal control, clustered around a Jupiter LOX tank; the paint scheme also helps visually track the booster's movements after launch. Its transporter is at right.

Cranes move the stage horizontally toward the service structure for placement on the launch stand.

The stage is moved into position as workers on the ground carefully control its movements with taglines. This booster would be refurbished and used to launch Apollo 5, an unmanned test flight of the LM in earth orbit, in 1968. One of the service structure's four small twelve-wheel red "trucks" that move the structure on rails is visible at lower right.

The stage is almost seated on the launch stand.

NAA technicians inspect the Apollo I CM on August 29 during checkout at the MSOB on Merritt Island. It is suspended from a bridge crane by its LES tower attachment wells.

A blue portable temperature controller covers the hatch as technicians work at the CM-SM umbilical, which carries the wiring and tubing for power, water, oxygen, and water-glycol flows between the modules. The connections are under an aluminum fairing protected by a red "remove before flight" cover and will be severed by a guillotine for CM separation before reentry. Other covers protect RCS thrusters and windows.

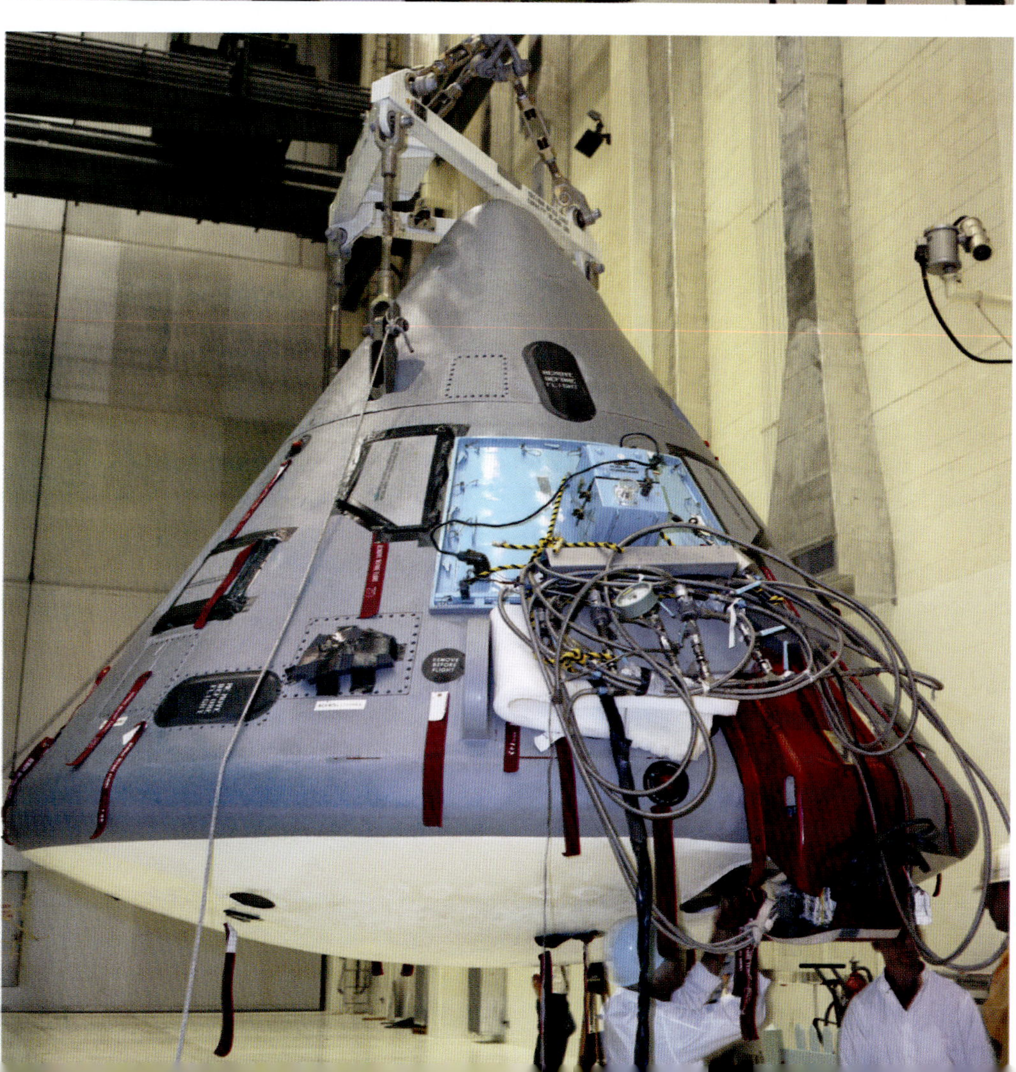

This view from below the CM shows the three oval-shaped attach points on the edge of the heat shield for the 4-inch-long stainless-steel straps that will connect it to the SM. They will be severed by small explosive charges for reentry.

A bridge crane in the high bay moves the CM toward altitude chamber L for mating with the SM on September 2.

★ **The two KSC altitude chambers were designed and manufactured by the Stokes Equipment Division of the Pennsalt Chemical Corp. of Philadelphia, Pennsylvania, and were installed in 1965 in the MSOB's high bay. All CSM tests used chamber L, a mirror twin of chamber R, used for lunar module tests. Both chambers are about 33 feet in diameter and 59 feet high, with walls of half-inch-thick stainless steel. They were first used in late 1965 to test CSM-009 for the AS-201 mission, but they were not man-rated until 1966. Although capable of simulating an altitude of 250,000 feet above sea level, the altitude tests on Apollo spacecraft were conducted at a simulated altitude of 210,000 feet.**

Technicians monitor the move. The chamber's 27.5-ton domed lid has been lifted by the high bay crane; the lid's weight will press down on a gasket to form the chamber's upper seal.

The CM is lowered into chamber L for mating with the SM for the first time. A cameraman, *at left*, lying on the closed chamber R lid films the move.

The Saturn IB's S-IVB-204 second stage is prepared for mating at LC-34 on August 31, with its yellow handling kit attached. Its single Rocketdyne J-2 liquid hydrogen engine is still wrapped at left.

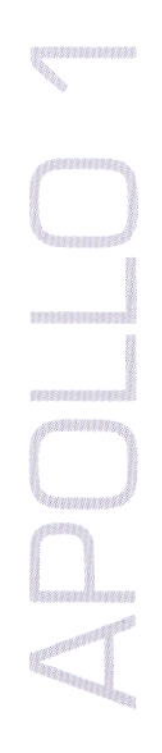

The stage rests vertically on the open legs of its handling kit. The interstage is partially visible at left.

The 54-foot S-IVB stage is raised, showing its J-2 engine. The aft interstage will be attached next.

The S-IVB aft interstage has been attached and the second stage is raised into position on August 31. The interstage will be bolted to the first stage and will remain with it at separation, about two and a half minutes after launch.

The S-IVB is lowered for mating onto the first stage. Work platforms line both sides of the service structure.

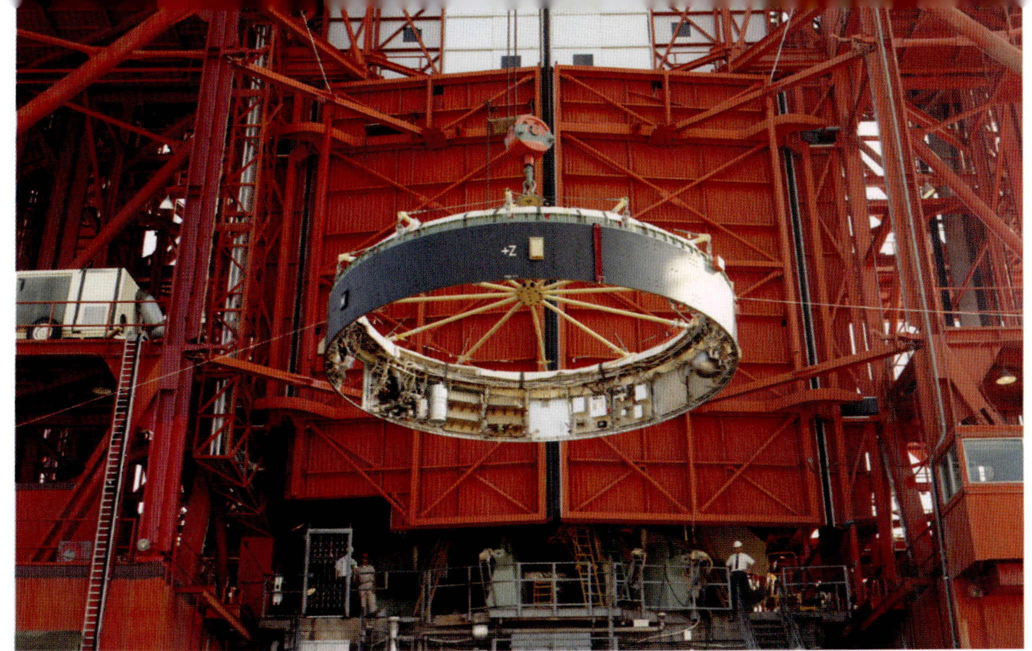

The IU, nearly 22 feet in diameter and 3 feet tall, is raised on August 31. Besides serving as the booster's "brain," the unpressurized IU also includes batteries, power supplies, and equipment for heat dissipation.

The IU is moved into position above the S-IVB.

Technicians guide the IU the final few inches for mating with the S-IVB's forward skirt. Despite having erected a nearly complete launch vehicle, NASA remains coy about its possible manned use as early as December.

Jim McDivitt, standing in the hatch, undergoes CM recovery training at Ellington AFB, southeast of MSC, with fellow backup crewmen Dave Scott (*top*) and Rusty Schweickart in September 1966. Two months later, the three are reassigned after the order of missions is shuffled; and in December they are named as the prime crew for an Earth-orbit test of the LM

The CM couches for Apollo 1 await installation in the MSOB on September 29. Manufactured by Weber Aircraft Division of Walter Kidde and Co. of Burbank, California (which produced the Gemini ejection seats), the couches are individually adjustable units made of hollow steel tubing covered with a fireproof Armalon fiberglass cloth. They are each supported by a head beam and two side beams, which are supported by eight struts to absorb the g-forces of ascent and splashdown.

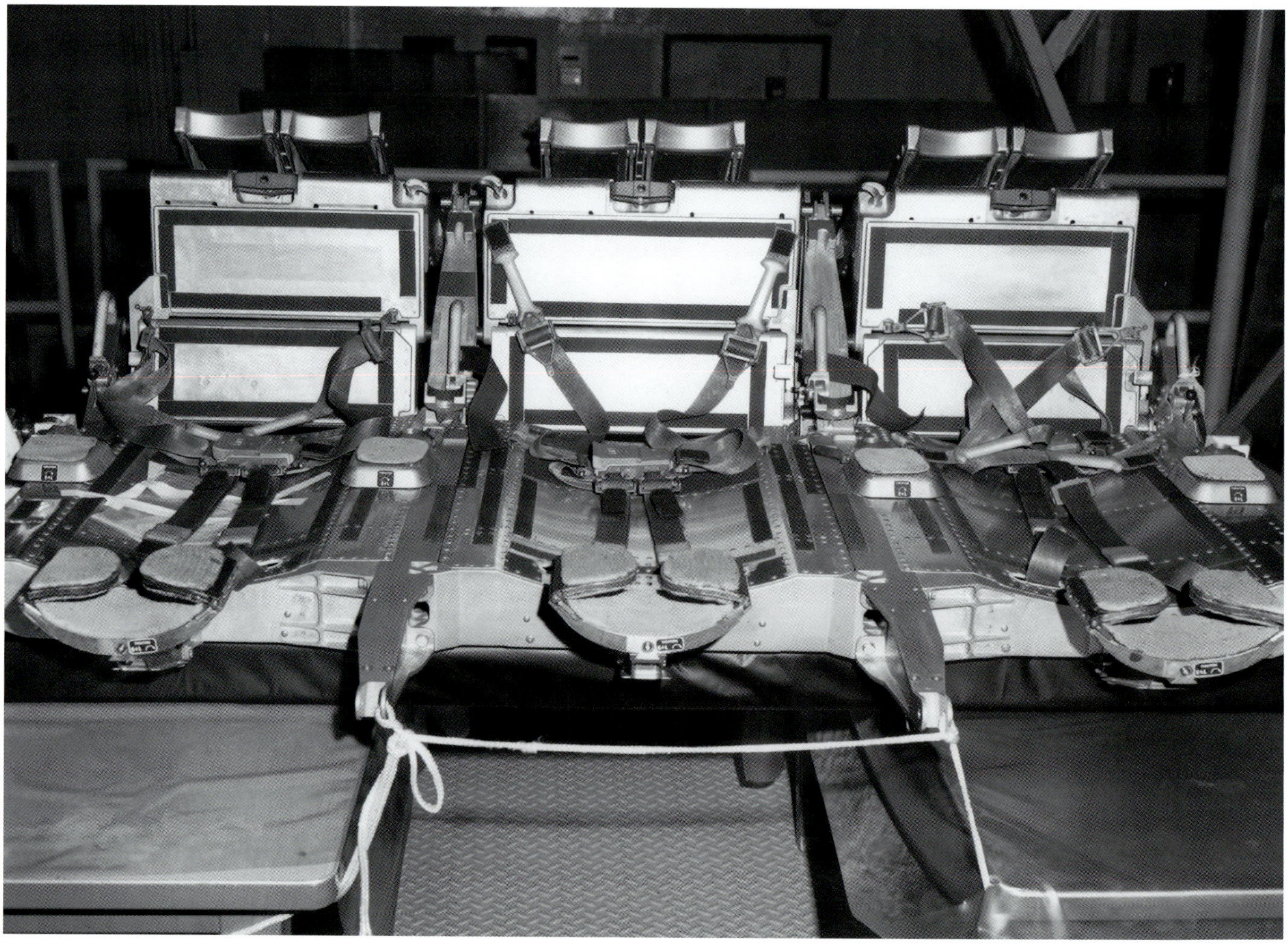

The crew for Apollo 2, command pilot Wally Schirra (*front*), senior pilot Donn Eisele (*right*), and pilot Walt Cunningham, is named on September 29. Their flight is intended to be a repeat of Apollo 1, with more emphasis on science. The crewmen pose the day before at MSC in front of the first Apollo Mission Simulator (AMS), built by the Link Group of General Precision Systems of Sunnyvale, California. The second AMS had been delivered to KSC that spring.

CHAPTER 6

October 1966

Swimmers steady CM-007 in the Gulf of Mexico off Galveston, Texas, on October 2 to qualify the spacecraft for postlanding operations. Aboard during the forty-eight-hour flotation test are MSC Flight Crew Support Division engineers Tex Ward and Lou DeWulf, and Harry Clancy from the Landing and Recovery Division. They report that the CM was more stable in the water than a Gemini reentry module. NASA Motor Vessel (MV) *Retriever* remained nearby.

Crew and support personnel pose aboard MV *Retriever* after the test. *Left to right*: test conductor Jim Shannon; crewmen Ward, DeWulf, and Clancy; test conductor Larry Bell; and, in the hatch, C. Mac Jones, vehicle project engineer. MV *Retriever* is a World War II–era Landing Craft Utility transferred to NASA from the US Army. It was used to train astronauts for postsplashdown ocean recovery operations during the Gemini and Apollo programs from 1963 to 1972. It operated primarily in Galveston Bay and the Gulf of Mexico.

Divers assist an "incapacitated" crewman during emergency egress training with CM-007 at the Ellington AFB pool on October 5.

Engineers from the Landing and Recovery Division of MSC wear Apollo 1–type space suits during training at Ellington on October 5. *Left to right*: Frank Janes, Paul Kruppenbacher, and John Hirasaki.

During October, the Apollo 1 CSM is put through a series of unmanned and manned altitude chamber tests in chamber L at the MSOB.

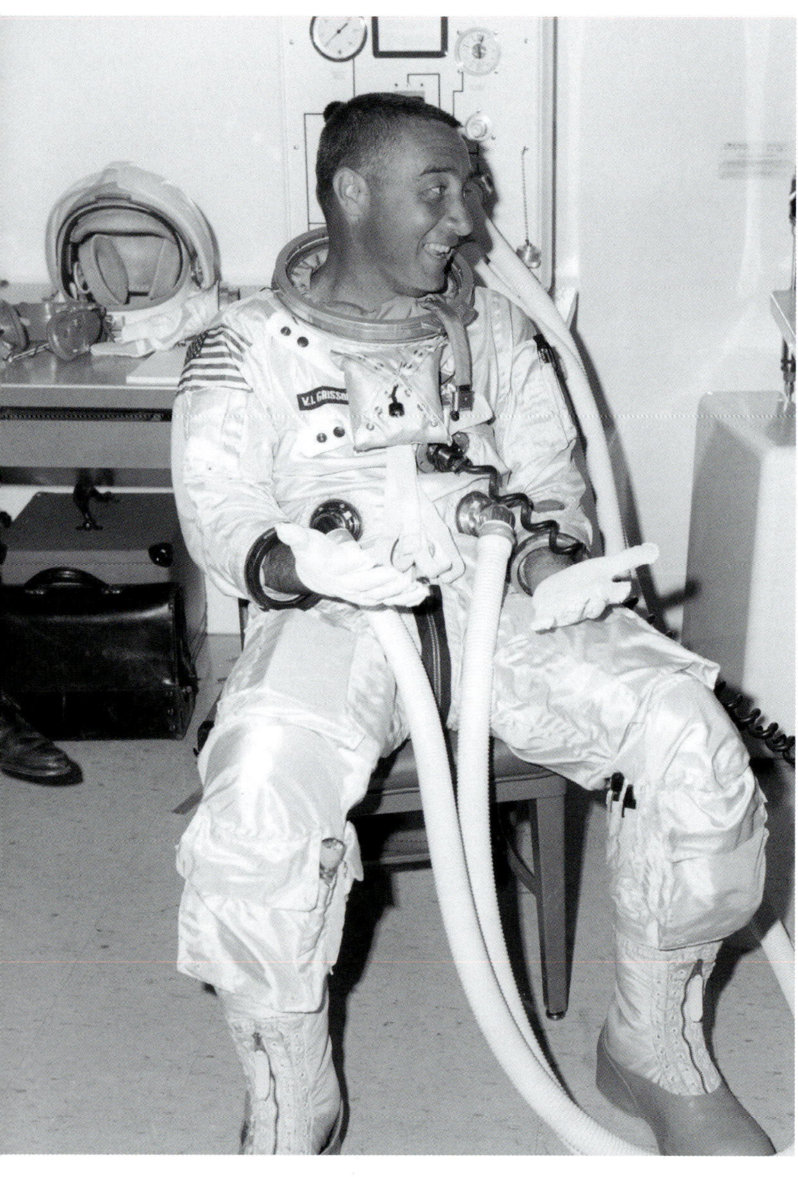

On October 13, Grissom is suited in an emergency clinic adjacent to the altitude chamber, with his headset and helmet behind him. NASA had decided to use the G3C Gemini suit as the base for the Apollo Block I suit, designated A1C. Its connectors were modified to accept Apollo hoses, and a protective shell was added over the Plexiglas helmet visor.

A coiled cable connects to White's biomedical and communications port; just above it, a pouch contains an inflatable water wing. His shoulder and leg pockets carry pens and flashlights.

The hose on Chaffee's right side is the inlet for oxygen; the hose at left is the oxygen outlet.

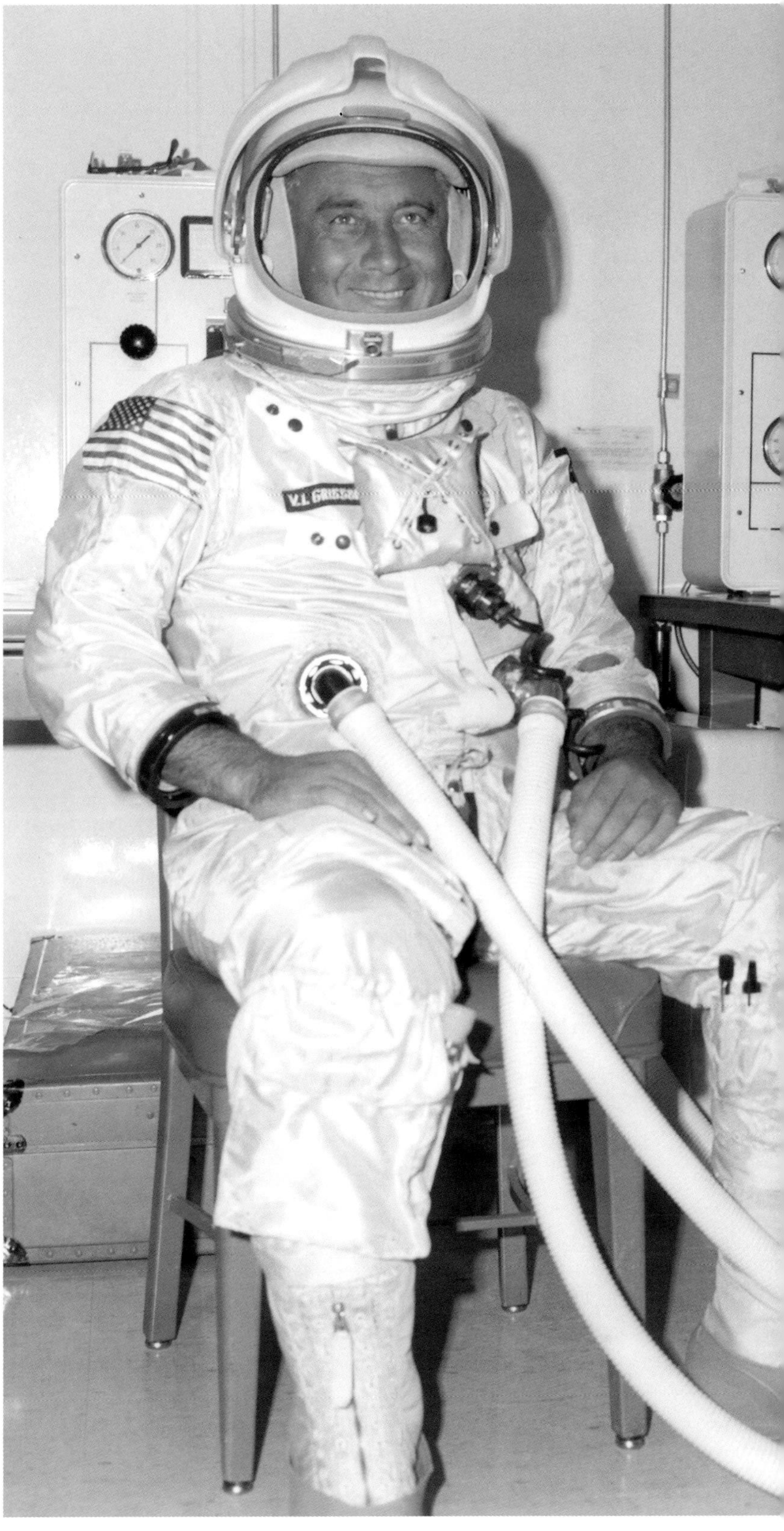

Grissom is fully suited. Each astronaut is connected to a portable pressurization unit; future Apollo crews would use a permanent suit room on the third floor of the MSOB.

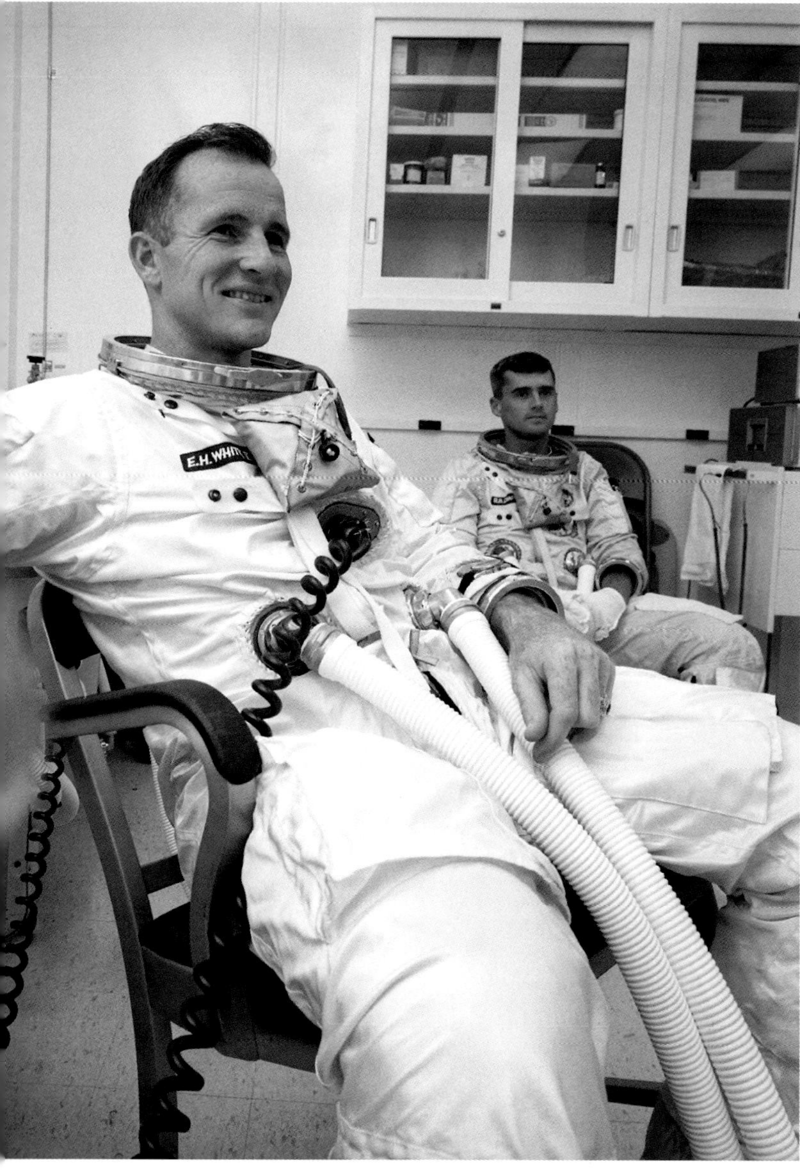

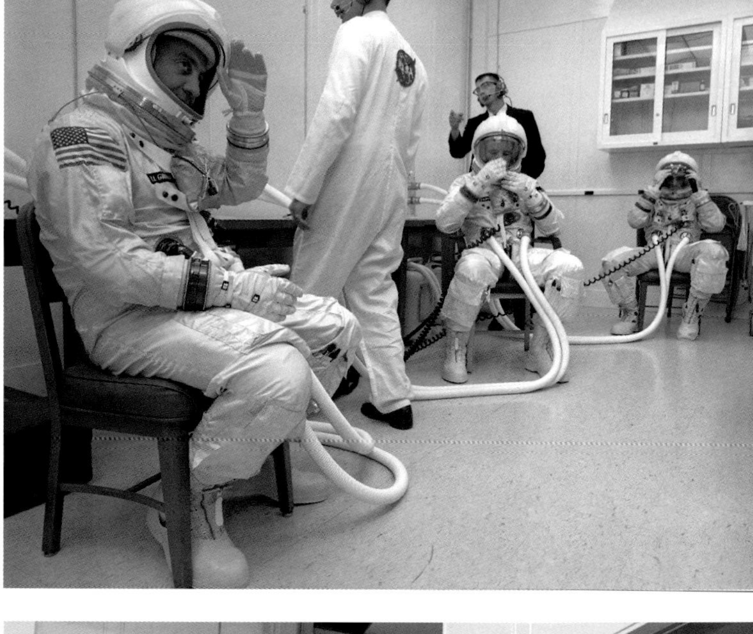

White (*left*) and Chaffee

The three crewmen close their visors.

Backup crewmen (*left to right*) McDivitt, Scott, and Schweickart watch as the prime crew prepares to head to the altitude chamber. They had spent seven hours overnight inside the CM, checking controls to prepare the spacecraft for the prime crew.

Grissom leads his crewmen toward the chamber, each still connected to portable suit ventilators.

Grissom (*partially hidden at left*), White, and Chaffee depart the clinic.

Grissom enters the airlock to the chamber, followed by White and Chaffee.

Grissom enters Apollo I as White (*left*) and Chaffee (*behind Grissom*) stand by. NAA's Don Babbitt, who would be overcome by smoke while trying to rescue the astronauts after the fire, is at center wearing headphones. A circular blowout plug at the tip of the CM is for a pilot parachute to help carry away the apex cover after reentry.

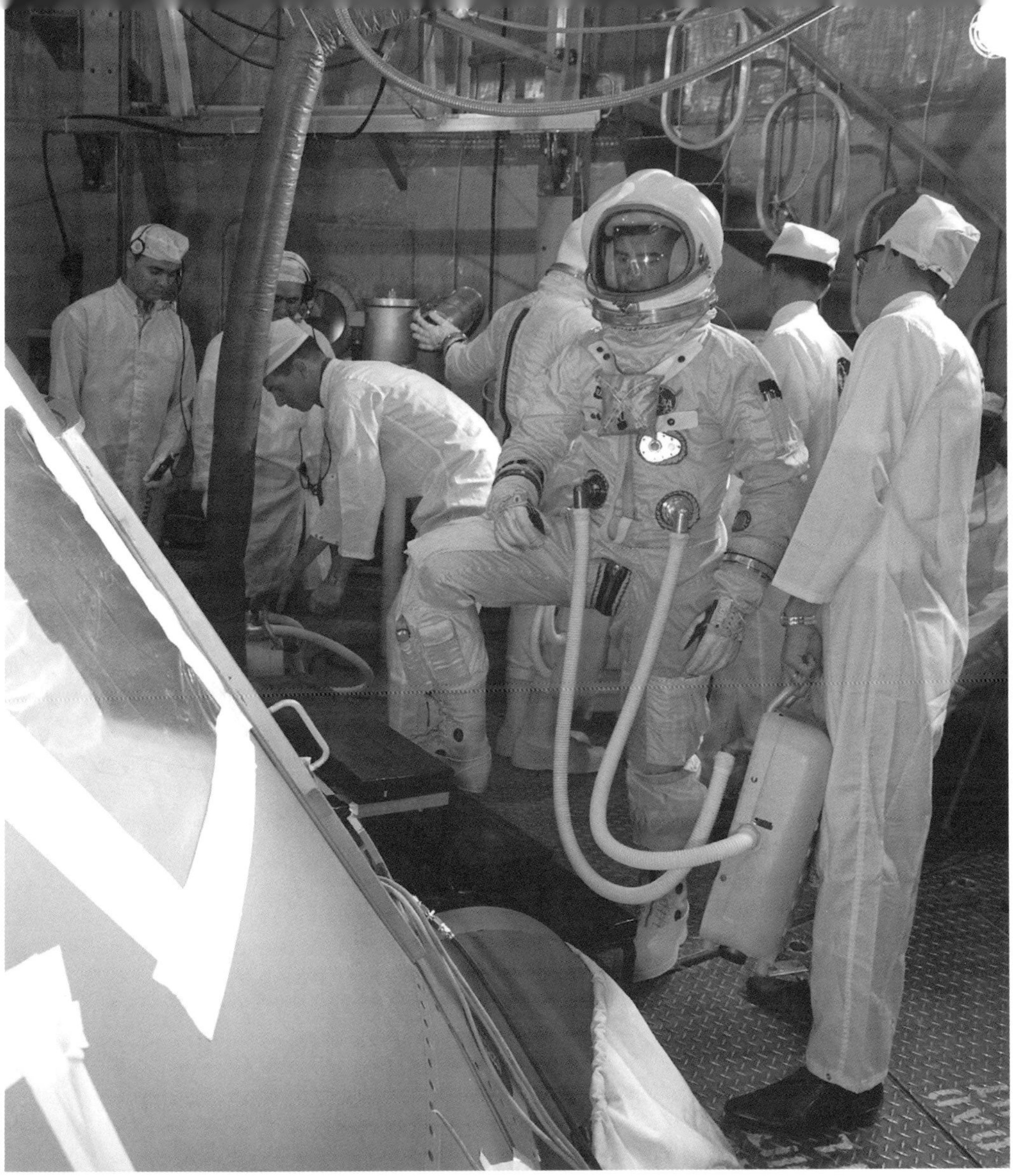

Chaffee waits to ingress.

White (*left*), who will occupy the center couch, will be the last to board. Chaffee prepares to ingress in the background.

A technician removes Chaffee's protective plastic boot covers.

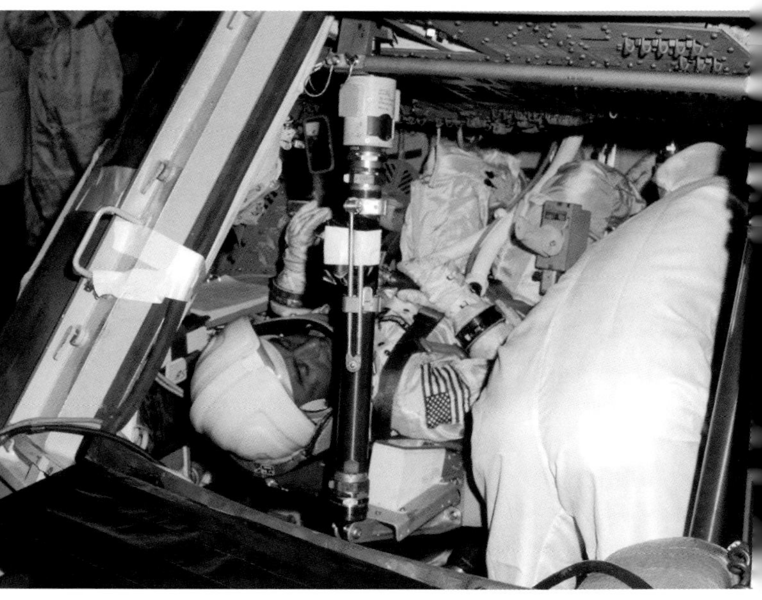

As commander, Grissom gets settled in the left-hand couch as a technician helps Chaffee (*not seen*) strap in about 8:00 a.m.

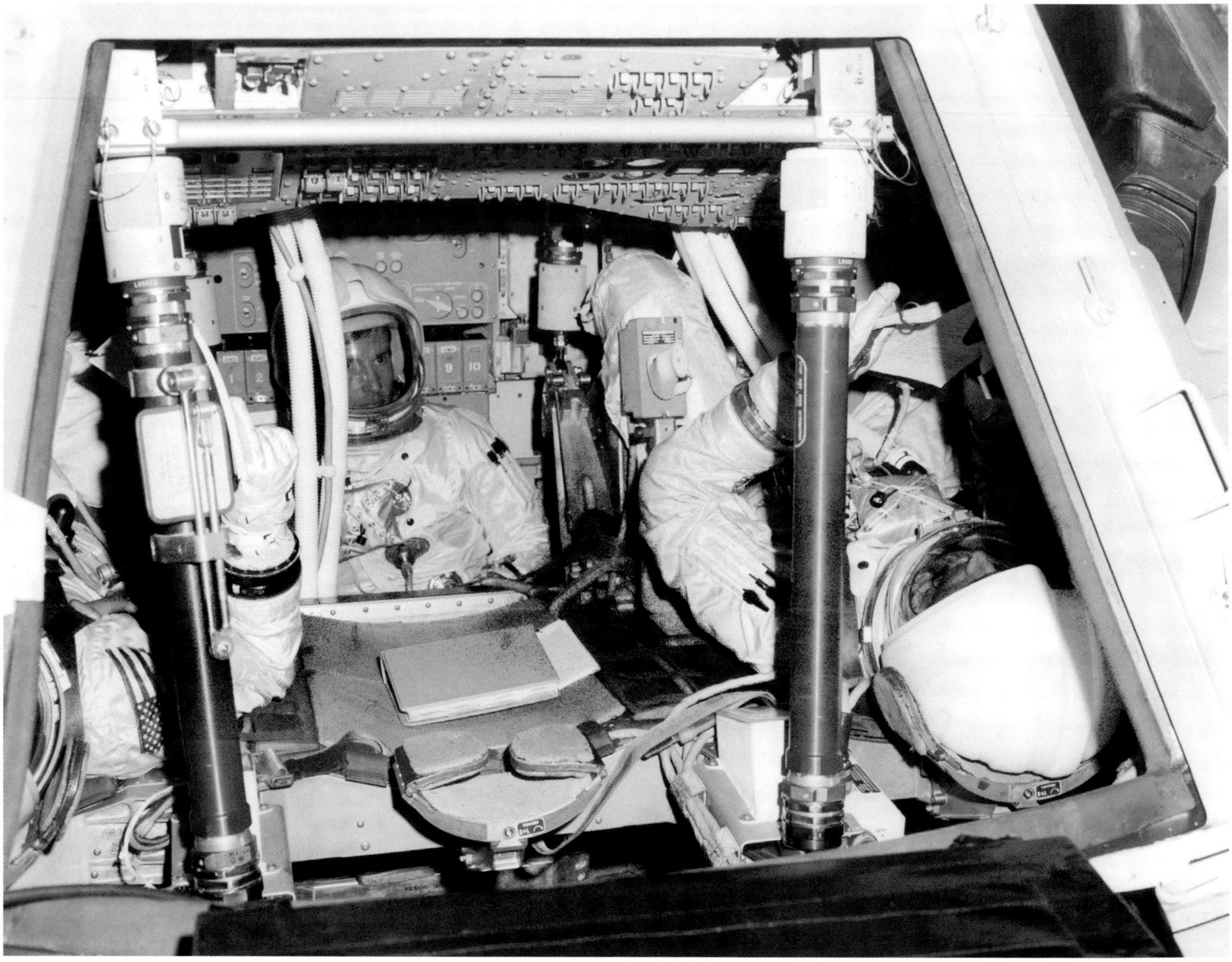

White occupies the lower equipment bay as Chaffee (*right*) checks paperwork. This more-than-six-hour systems check is conducted at sea-level atmospheric pressure in the chamber.

October 15:
Unmanned Altitude Chamber Run

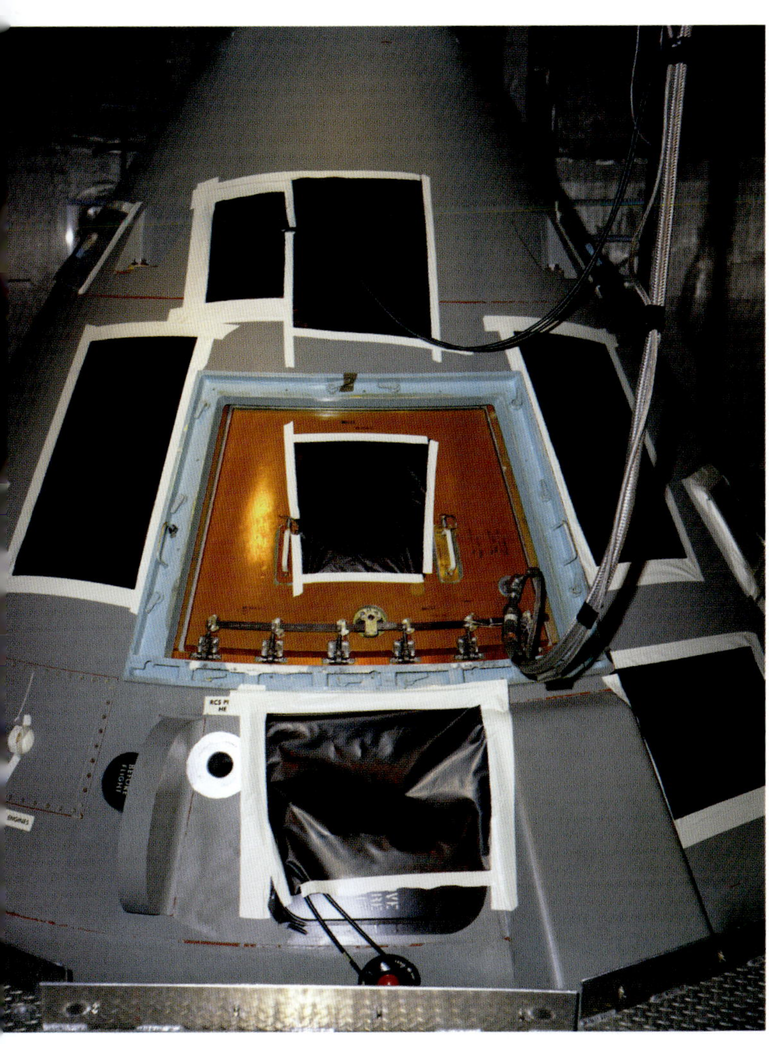

Apollo I's gold-tone inner hatch, a removable, lightweight structure with interlocking edge members on three sides, is in place before the test. The bottom is secured by six latches attached to a rack-drive bar. Black coverings protect the CM's windows and RCS engines. Handholds are on both sides of the center window. Cabin pressure is used to aid in sealing.

Side view of Apollo I. This test will be conducted at altitude pressures.

Cables and hoses from the SM's umbilical area just below the CM are connected to chamber test equipment.

An SM access panel with an RCS thruster quad is not yet closed before the test.

Two of the SM's fuel cell power plants occupy a shelf; the third fuel cell is behind them. They provide most of the spacecraft's electrical power and some of its drinking water. Their oxygen and hydrogen supply tanks are below them.

Each covered fuel cell is 44 inches high and 22 inches in diameter and weighs 245 pounds.

October 18–19: Manned Altitude Chamber Runs

Grissom prepares for a chamber run that will simulate an orbital altitude. Although recliners have been added, the clinic still serves as the temporary suit room. Motion picture photographer Larry Summers, working for NASA contractor Technicolor, films with a 16 mm Arriflex camera at right.

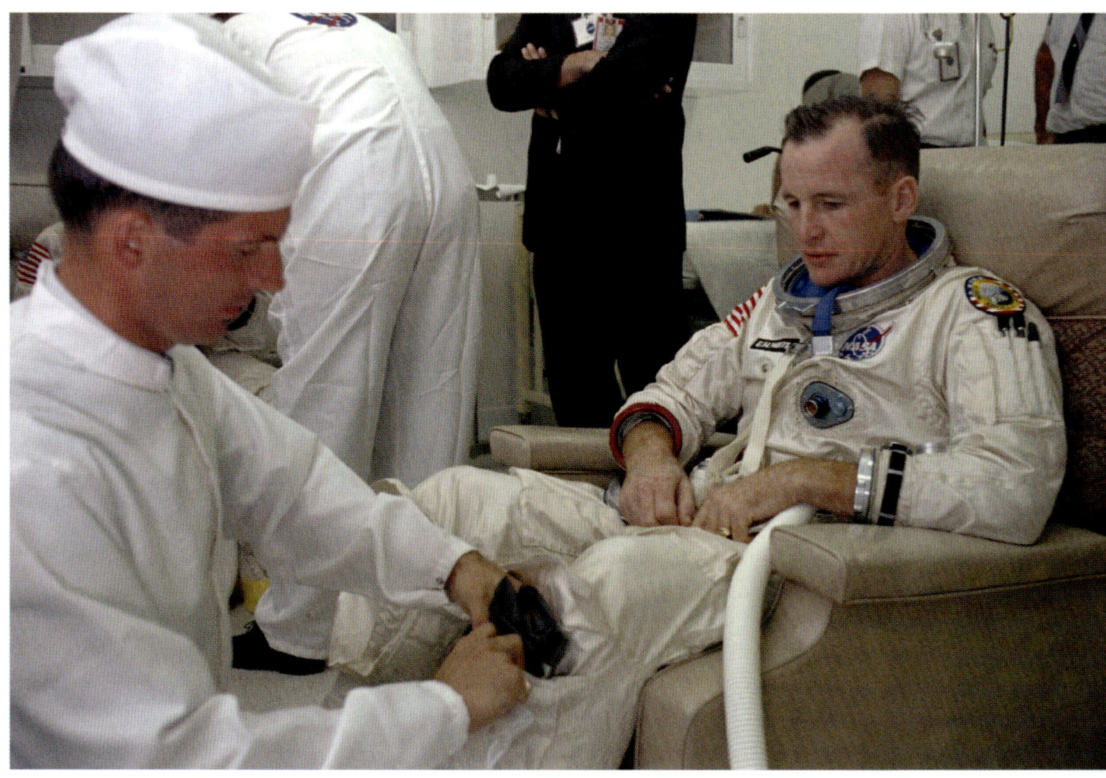

Suit tech Joe Rebokus stows an item in White's shin pocket. The crewmen wear an embroidered version of their mission emblem for the first time, here on the right shoulder.

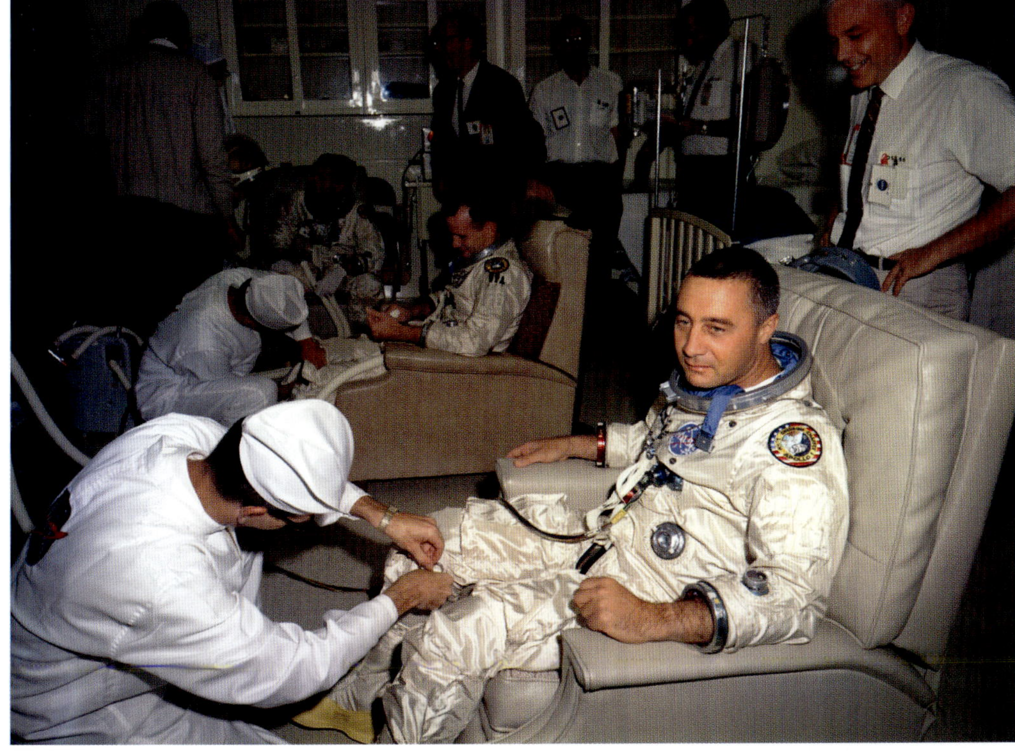

Activity picks up as technicians work on all three crewmen's suits.

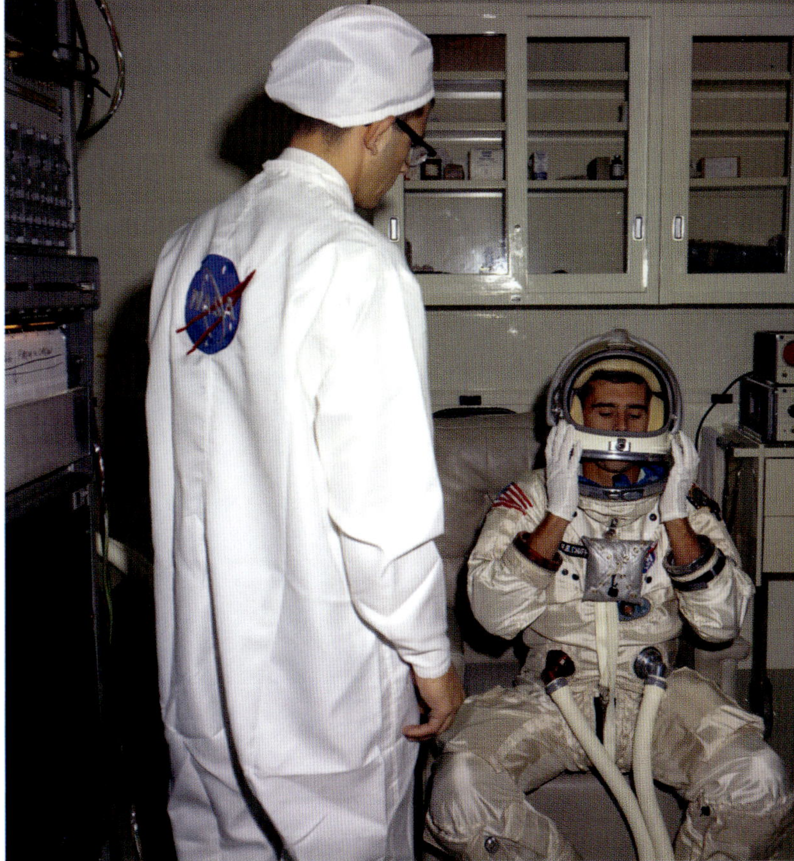

Chaffee dons his helmet. The chart recorder at left is ready for the crew.

The Apollo I crew emblem was designed by NAA graphic artist Allen Stevens. During the 1960s and 1970s, he worked for the successor companies of North American Rockwell and Rockwell International before retiring in 1978. He said he spoke to the prime crew only two or three times. Although he initially proposed a phoenix (rising from the flames) theme for Apollo 7, NASA rejected it as possibly being in bad taste. Stevens would go on to design emblems for the remaining Apollo missions, but only his designs for Apollos 7, 9, and 10 were officially used. The embroidered versions worn by the crew on their pressure suits were the last ones approved, banned after the fire because of their flammability.

Two preliminary versions of the Apollo I crew emblem by NAA graphic designer Allen Stevens.

A suit technician snaps a strap on Grissom's suit.

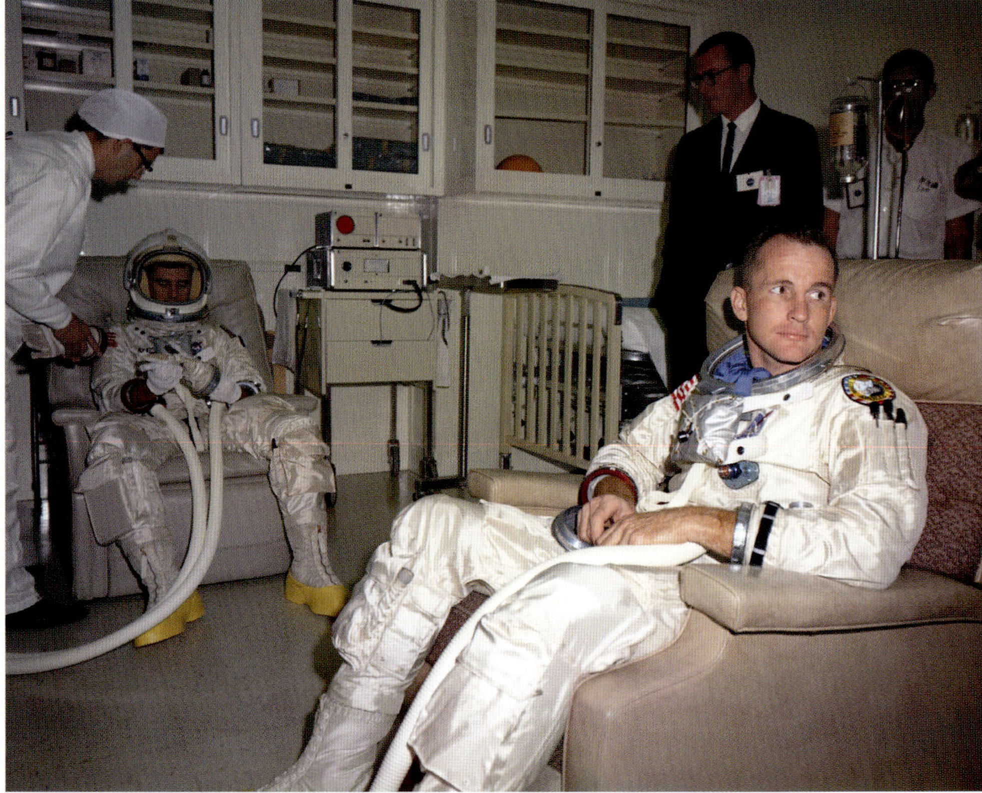

Technician Joe Rebokus checks White's water wings, with Technicolor photographer Larry Summers in the background.

Chaffee (*left*) and White hold gloves.

Grissom adjusts his helmet.

Chaffee waits for a leak check.

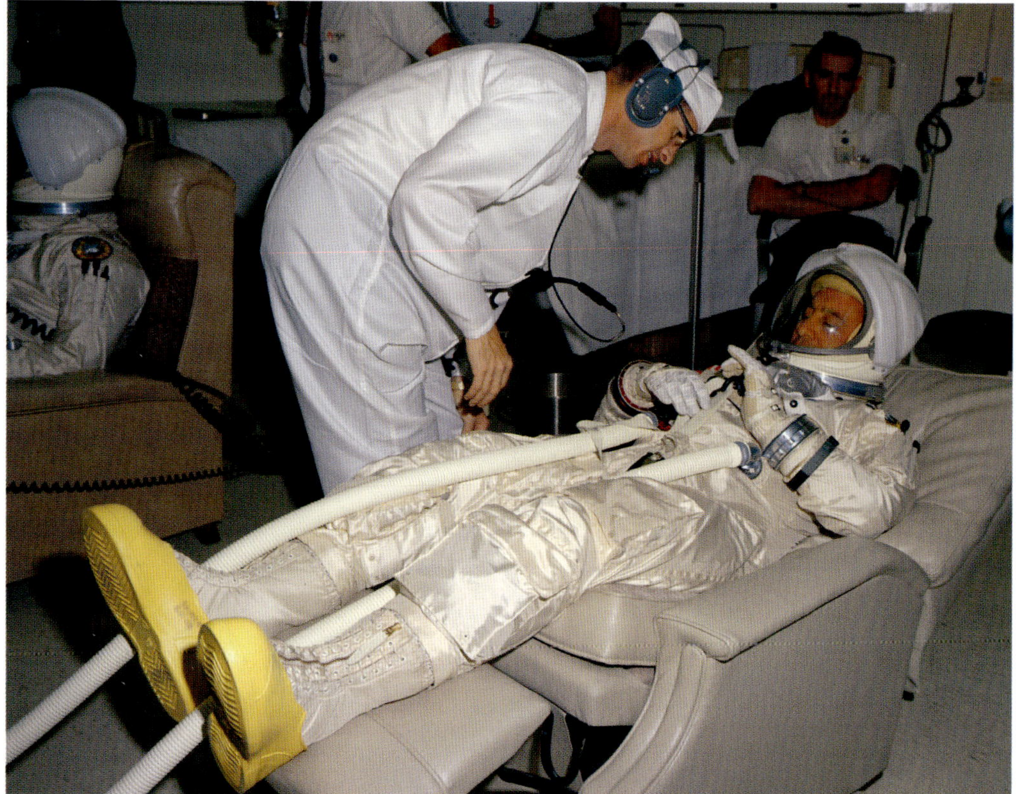

A suit technician checks communications with Grissom, with White partially visible at left.

Chaffee's suit undergoes a leak check. Summers films at right.

Grissom walks toward Apollo I. NAA technicians wear blue smocks.

Grissom begins to ingress. The black covers taped over the windows have been removed.

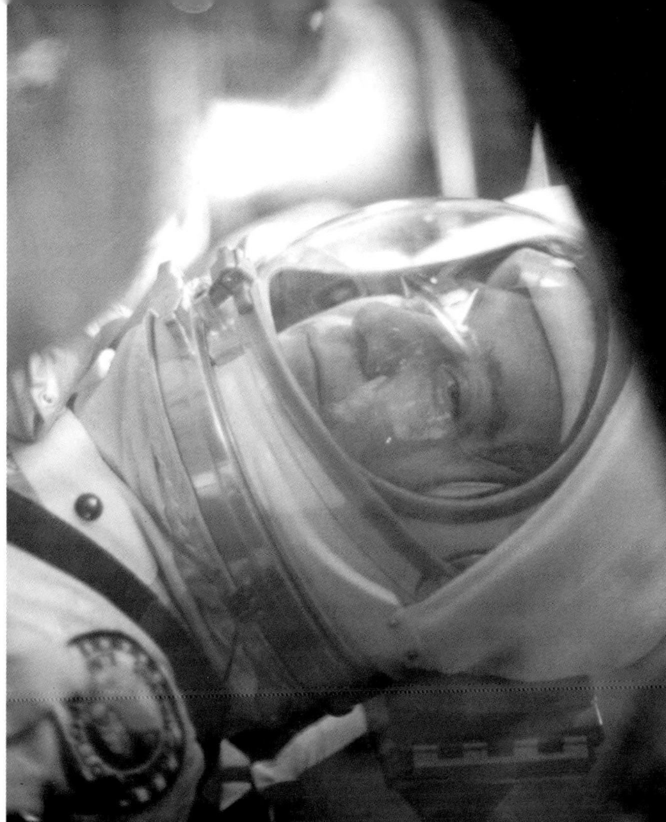

Grissom, seen through side window no. 1, gets settled in the commander's couch.

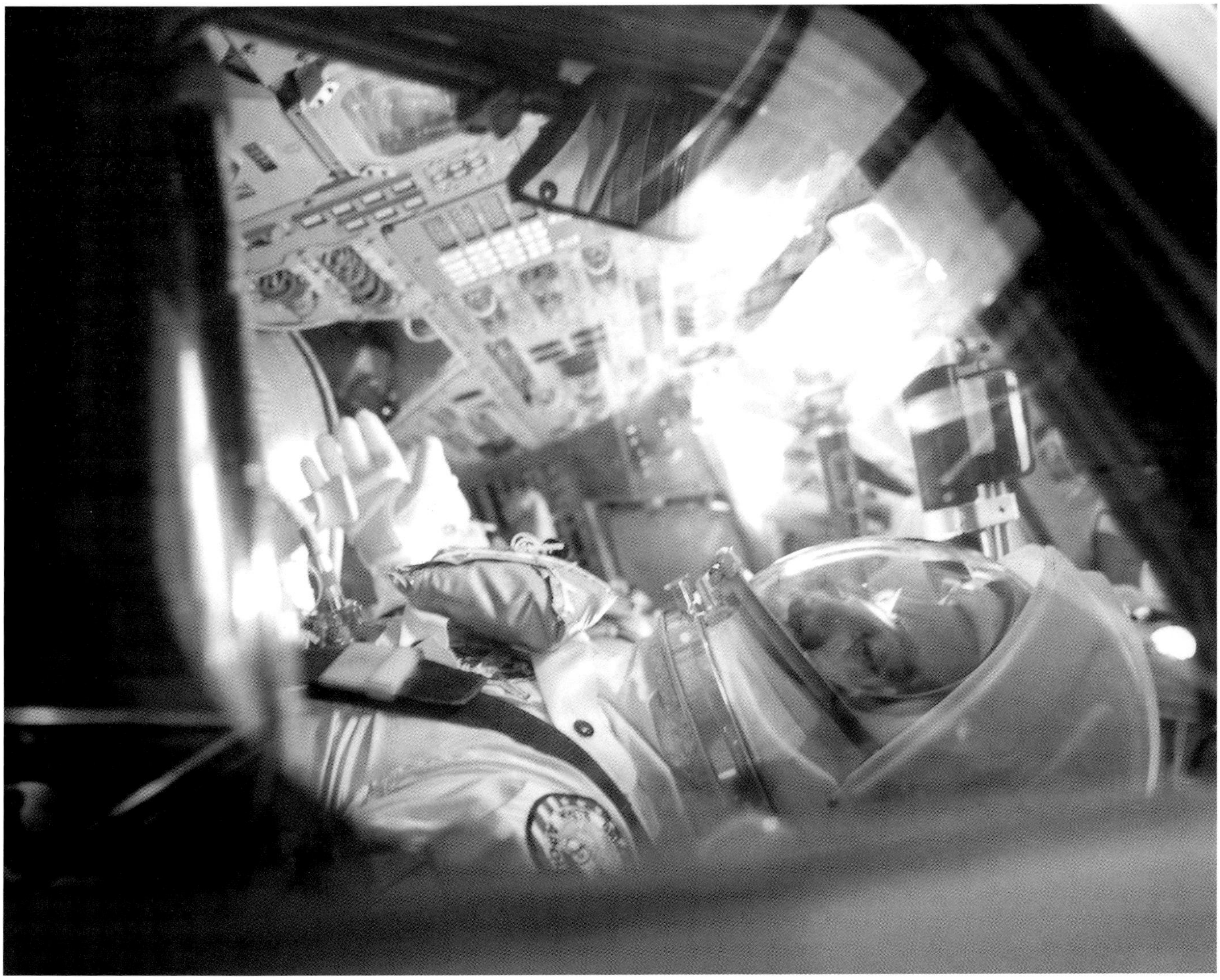

Chaffee will ingress next, occupying the right-hand pilot's couch.

A technician makes final adjustments for Chaffee.

Senior pilot White will be the final crewman to enter.

White with technician Joe Rebokus (*left*)

Grissom inside Apollo I, seen through rendezvous window no. 2 and side window no. 1.

The crew is in place. Latches to secure the inner hatch surround the opening. Personnel from the Launch Support Division of Bendix Corp. are responsible for chamber operations, and a three-man Bendix rescue team is on standby in the airlock.

Technicians install the spacecraft inner hatch before the main vacuum pump begins to bring the chamber to altitude. The test is stopped, however, after reaching a simulated altitude of 13,000 feet when a CM electrical component fails. A repressurization system returns the chamber and airlocks to sea level by gradually introducing breathable air.

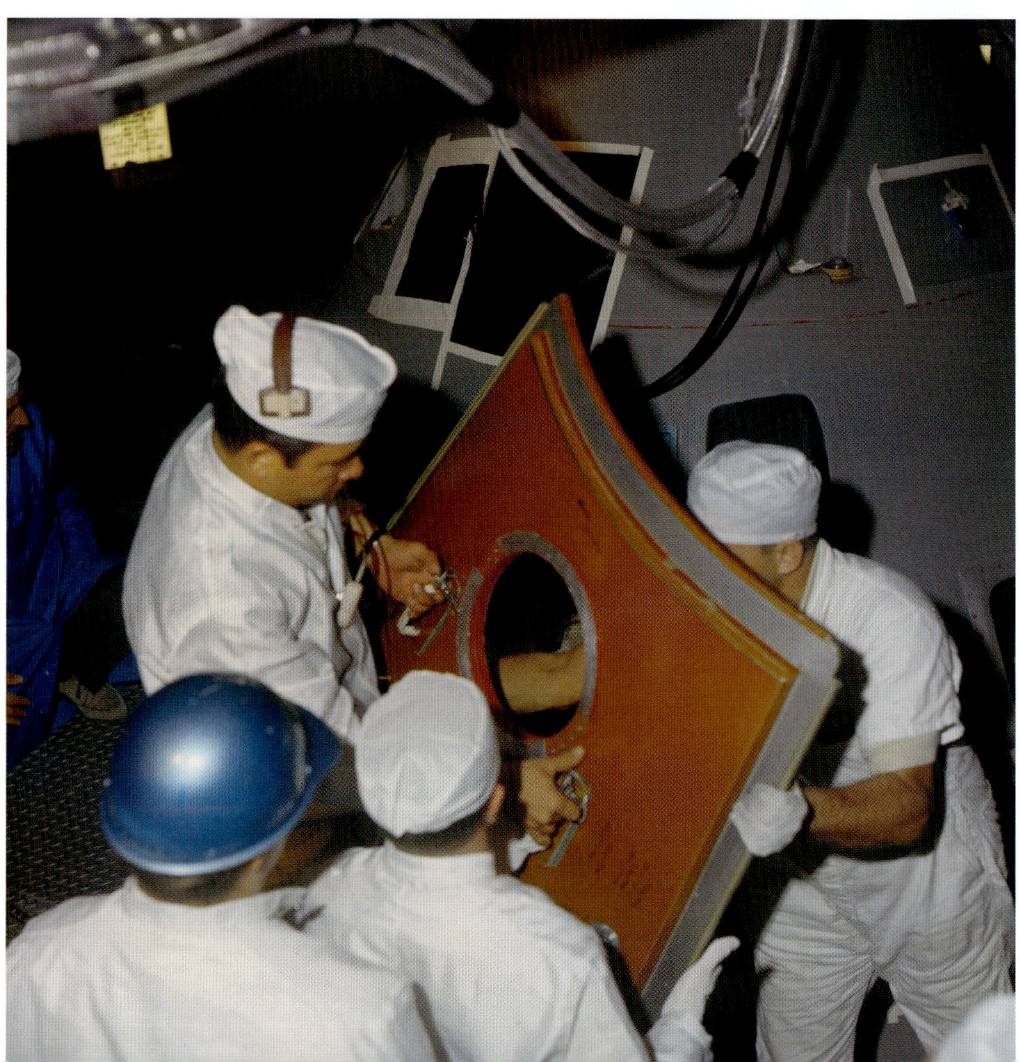

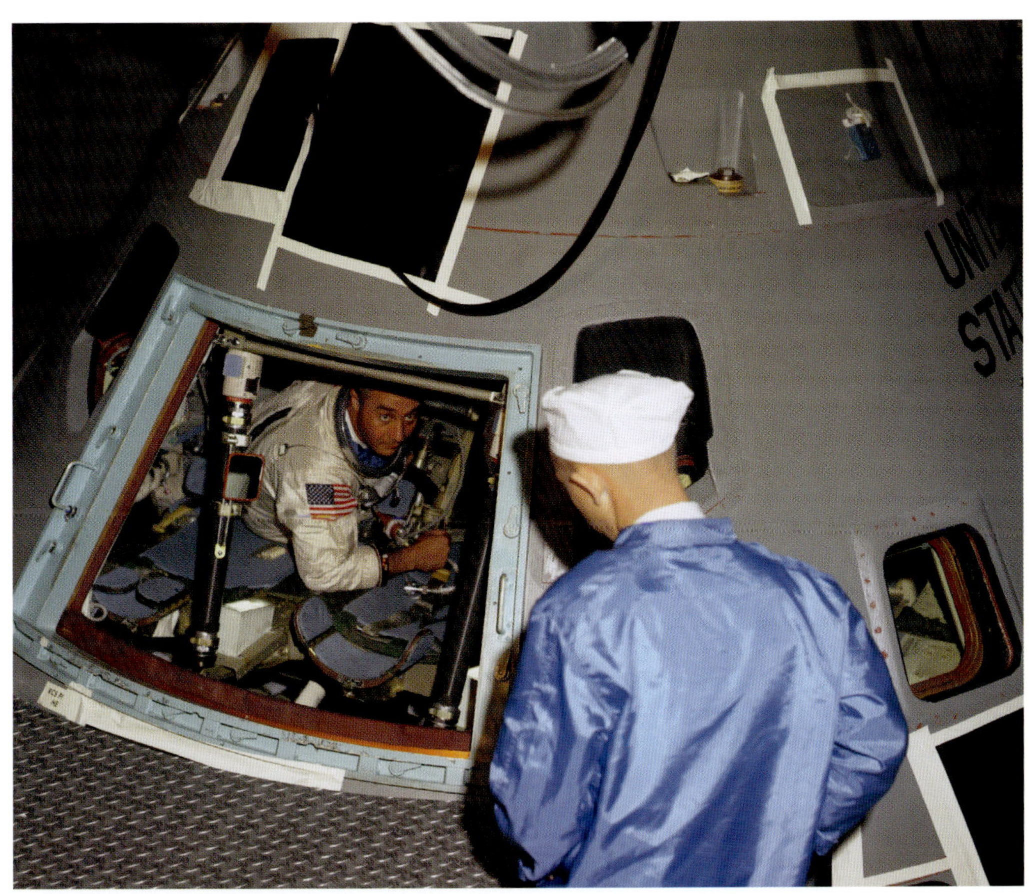

Grissom is the first to egress after the abbreviated chamber test. The astronauts would successfully repeat the sixteen-hour test the next day. A planned run with the backup crew on October 21 is postponed before the crew can board after a failure in an oxygen system regulator in the spacecraft's Environmental Control Unit.

The prime and backup crews for Apollo 2 pose with MIT engineers on October 19 during navigation software training at the MIT Instrumentation Laboratory in Cambridge, Massachusetts, which designed the Apollo Guidance Computer. *Front row, left to right*: prime crew Wally Schirra, Walt Cunningham, and Donn Eisele; backup crew Tom Stafford, Mike Collins, and Frank Borman. *Second row, left to right*: Russel Larson, Balraj Socappa, Wayne Templeman, Jerry Levine, Duson Alexander Koso, James Nevins, and David Hoag. *Back row, left to right*: Ralph Ragan, Malcolm Johnston, Jack Dunbar, Jack Schellingfurd, unidentified, John Dunbar, Dan Lickly, and Ivan Johnson. A diagram of a Block I CM Display Keyboard is on the easel at left. *MIT photo*

The Apollo 2 prime crewmen are suited at NAA in Downey to try out the couches in CM-014 on October 25. Their spacecraft is several months behind Apollo 1 in the processing flow. *Left to right*: Eisele, Schirra, and Cunningham.

The Apollo 2 astronauts aboard CM-014. Command pilot Schirra is out of view at left, senior pilot Eisele is in the center couch, and pilot Cunningham is on the right. A technician prepares to egress after helping them strap in.

On October 26–27, the Apollo 1 prime and backup crews participate in recovery training with Boilerplate 1101A, using MV *Retriever* in the Gulf of Mexico off Galveston, Texas.

On October 26, backup crewmen (*left to right*) Scott, McDivitt, and Schweickart prepare to board the boilerplate on the deck of *Retriever*.

★ **BP-1101A was an aluminum boilerplate** built to test the CM's flotation characteristics. It was fabricated at Kelly AFB in San Antonio, Texas, and was delivered to MSC as BP-1101 in early 1965 in a Block I configuration. It was returned to Kelly for modification to Block II as BP-1101A in late 1965. The boilerplate used internal ballast and exterior dummy equipment to approximate equipment location in the crew compartment. It was used to tests the Block I uprighting system and the flotation collar in MSC's Water Immersion Facility and in the Gulf of Mexico. The uprighting system was commanded by a hardwired box controlled by the test conductor from a life raft.

BP-1101A with the Apollo 1 backup crew aboard

Scott is hoisted to a waiting US Coast Guard Sikorsky HH-52 Seaguard helicopter as McDivitt waits in a raft. The chopper is based at Ellington AFB.

Grissom relaxes aboard *Retriever* the next day prior to more recovery training, wearing a windbreaker with a Pure Oil Co. patch.

Grissom and White relax aboard *Retriever*.

Chaffee aboard *Retriever*

Chaffee, Grissom, and White relax in a raft aboard *Retriever*.

White, Chaffee, and Grissom go over notes with Tex Ward (*second from left*) of MSC's Flight Crew Support Division before recovery training on October 27. Ward, chief of the Apollo egress training program, was one of the crew members of the forty-eight-hour flotation test aboard CM-007 on October 2.

Grissom leads Chaffee (*partially hidden*) as the crewmen emerge for recovery training. Ward is at left. NASA senior photographer Bill Taub (*with tie*) is at upper left.

Grissom boards the CM boilerplate.

White (*left*) and Chaffee prepare to ingress.

White waits to board.

White climbs aboard the boilerplate. MSC photographer Andrew "Pat" Patnesky is at right.

The CM boilerplate with the prime crew aboard is lowered into the Gulf of Mexico by the large davit crane from the deck of *Retriever*.

Chaffee, Grissom, and White float in a raft. In the background is *Duchess*, a yacht owned by La Porte, Texas, businessman Paul Barkley and hired as an observation boat for journalists covering the training.

CHAPTER 7

November–December 1966

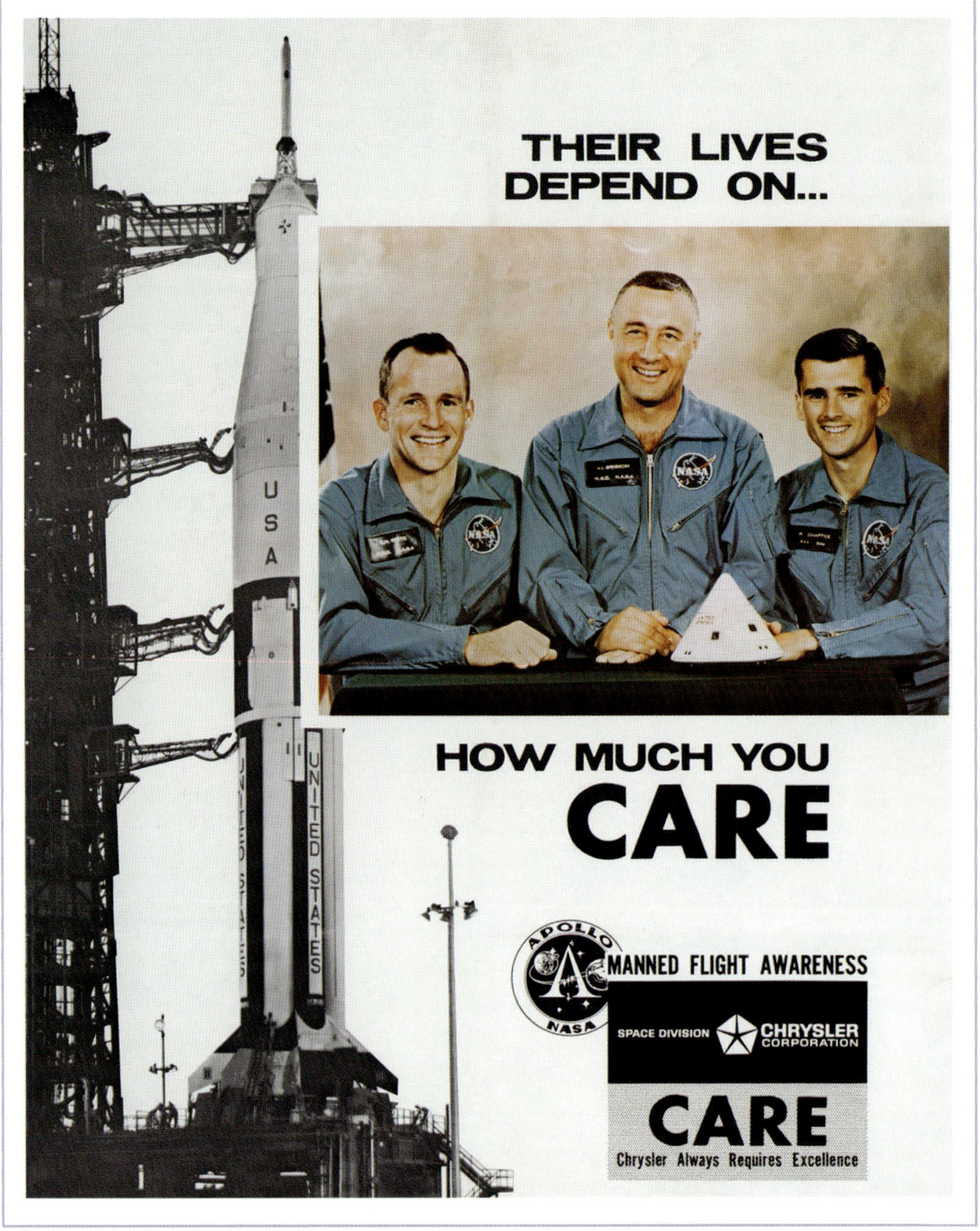

Chrysler's Space Division creates this poster for NASA's Manned Flight Awareness program.

Formalized in 1964, the Manned Flight Awareness (MFA) program was created to remind agency and contractor employees that lives were at stake in their daily work. Grissom, speaking to General Dynamics employees during a visit to an Atlas factory in 1959, simply told them, "Do good work," a phrase that was integrated into MFA messages. The program expanded after the Apollo I fire and produced posters, buttons, medallions, and other items. The MFA program also adopted the *Peanuts* comic strip character Snoopy as a safety mascot, and the Silver Snoopy award is presented by astronauts to NASA and contractor employees for an outstanding contribution to flight safety and mission success. During the Space Shuttle era, MFA was renamed Space Flight Awareness.

CSM-008 is in vacuum chamber A, the larger of two man-rated chambers in the Space Environment Simulation Laboratory in MSC's Building 32 in Houston. The spacecraft would undergo its second manned test near the end of October 1966, primarily to check out the Environmental Control System (ECS). Carbon-arc lamps mounted at right could simulate solar radiation on the spacecraft. Because the complete CSM was too tall to fit through the 39-foot door of the chamber, the CM and SM were moved in separately and stacked on top of the test platform. The huge chamber is 188 feet tall and 56 feet in diameter.

CSM-008 test crew members (*left to right*) Joe Kerwin and Ed Givens and USAF test pilot Joe Gagliano inspect the spacecraft. Kerwin is included as the only astronaut physician; Givens had been selected as an astronaut six months earlier. Gagliano was part of the backup crew of the first manned test in August. Rendezvous window no. 4 is in the foreground.

Kerwin (*left*) is the first to egress CSM-008 on November 1 at the end of the six-day test. Givens, also wearing a prototype headset, and Gagliano prepare to climb out. Some 550 NASA and contractor employees are involved in around-the-clock support of the test. The "flight" overcomes several problems, and both chamber tests lead to fourteen hardware changes.

The prime crew in the Apollo I's lower equipment bay in the MSOB on December 2. At right, White works with a T-handle hex wrench used to open and close access panels; Grissom (*at left*) holds another T-handle wrench.

Chaffee, Grissom, and White examine a mockup of the CM's Data Storage Equipment recorder in the MSOB on December 2.

The Environmental Control Unit (ECU) is removed from the CM on December 3 for repairs; several ECU units had been swapped out after they had repeatedly malfunctioned since October. Essentially the CM's air conditioner, it also manages the spacecraft's water supply and was manufactured by the Garrett Corp.'s AiResearch Division in Los Angeles, California. The ECU trouble and other problems push the revised launch date, tentatively scheduled for November, into the first quarter of 1967.

Left to right: Eisele, Schirra, and Cunningham chat before water egress training in the Building 260 water tank at MSC on December 5. On November 17, NASA had announced that the Apollo 2 mission would be canceled as an unnecessary repeat of Apollo 1. This Apollo 2 crew was then reassigned as the backup crew for Apollo 1.

Left to right: Astronauts Bill Anders, Frank Borman, and Mike Collins, who'd be named as the Apollo 3 crew a week later, pose with the S-IC-3, the first stage of their Saturn V, during a visit to the Michoud Assembly Facility in New Orleans, Louisiana. On December 22, the original Apollo 1 backup crew (McDivitt, Scott, and Schweickart) was also reassigned to a new Apollo 2 Earth-orbiting mission, to be the first flight of a Block II CSM with a separately launched lunar module.

MSC director of flight crew operations Deke Slayton also announces in December that a third crew would be added to each Apollo mission for additional ground support and training. The support crew for Apollo I would be (*left to right*) Group 5 astronauts Ed Givens, Jack Swigert, and Ron Evans.

New Apollo I backup commander Wally Schirra is aboard MV *Retriever* during recovery training in the Gulf of Mexico on December 6. Schirra's crew is several months behind the prime crew's schedule so has to catch up on recovery training the prime crew had completed in October.

Cunningham during recovery training

Eisele during recovery training

Apollo program manager Joe Shea speaks during the Apollo 1 preflight news conference in the Building 2 auditorium at MSC on December 16. A cutaway CSM model and an LES model frame Shea.

Shea briefs reporters, with public affairs chief Paul Haney seated at right.

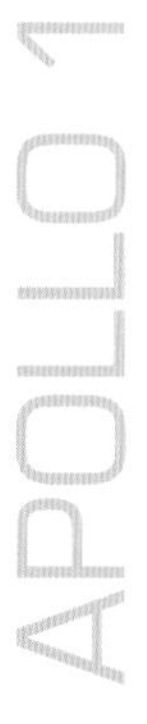

The conference participants include (*left to right*) Slayton; backup crewmen Cunningham, Eisele, and Schirra; and prime crewmen Chaffee, White, and Grissom. When asked, Schirra says he was disappointed that his crew had become backups, but he agreed with the move.

Left to right: **Chaffee, White, and Grissom wait to take questions.**

MSC public affairs officer Jack Riley (*right*) calls on a reporter, with Slayton at left holding his signature cigar. Shea announces that a TV camera would be aboard for the first time, possibly showing the crewmen removing their pressure suits. When a reporter jokingly calls that a striptease, Grissom replies, "I'm very bashful on camera. Would you believe [just] helmets and gloves?"

Grissom, White, and Chaffee pose next to the CSM model. Despite the problems that had meant delays, Grissom says, "The spacecraft systems all look pretty good to us now."

Chaffee talks with ABC News science editor Jules Bergman after the news conference.

Schirra dons his helmet before the backup crew takes part in the final Apollo I chamber test on December 29. It was planned for October but was postponed because of the continuing ECU problems.

Cunningham adjusts a glove before the test.

Suit tech Joe Rebokus helps Eisele with his neck ring.

Left to right: Eisele, Cunningham, and Schirra await leak checks.

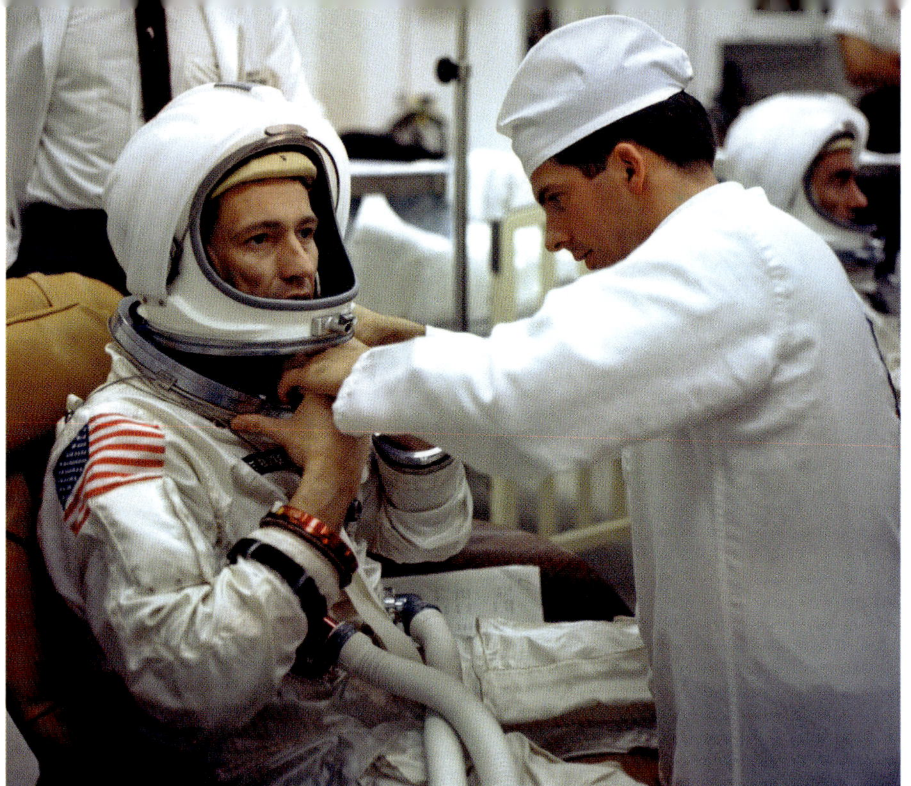

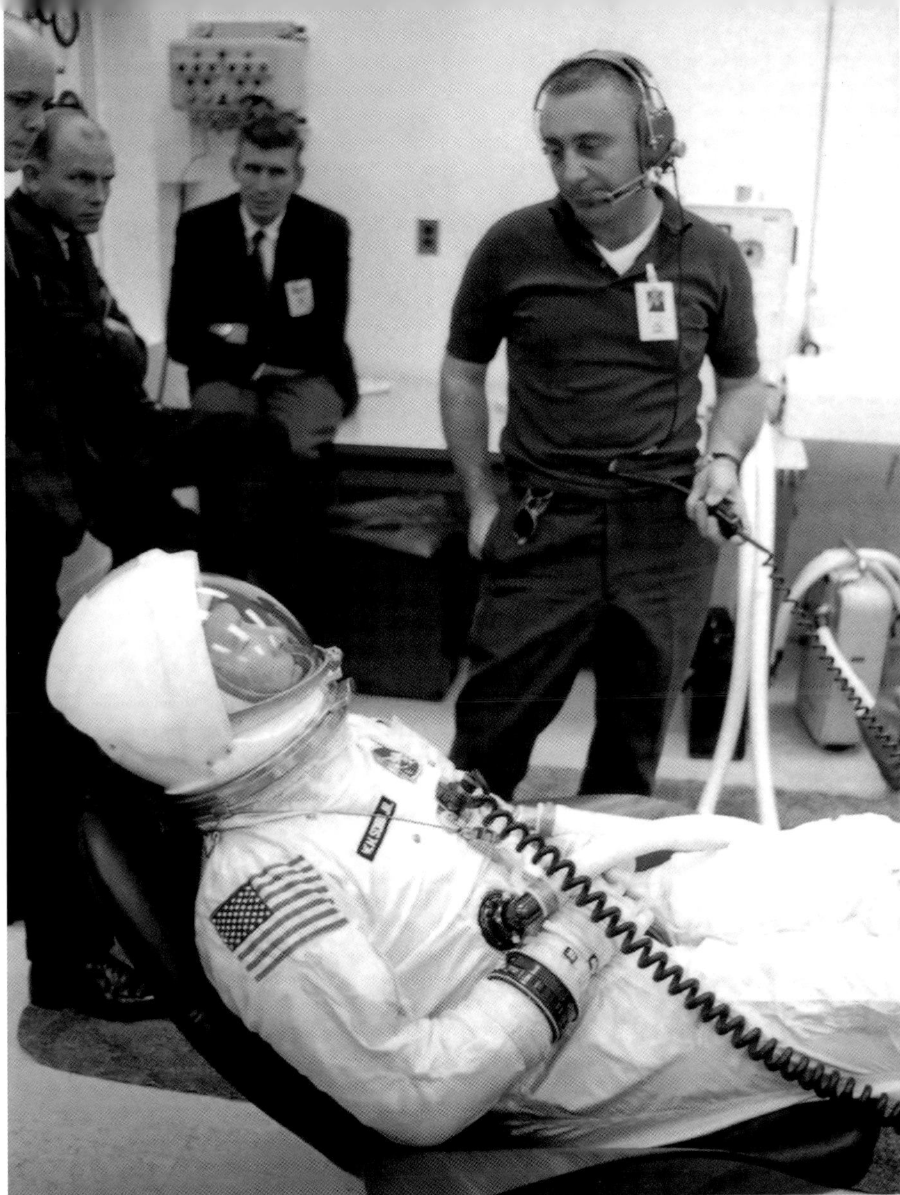

Grissom (*right*) talks with Schirra.

Eisele looks over a NASA *Spacecraft Operations Checkout Procedure* manual.

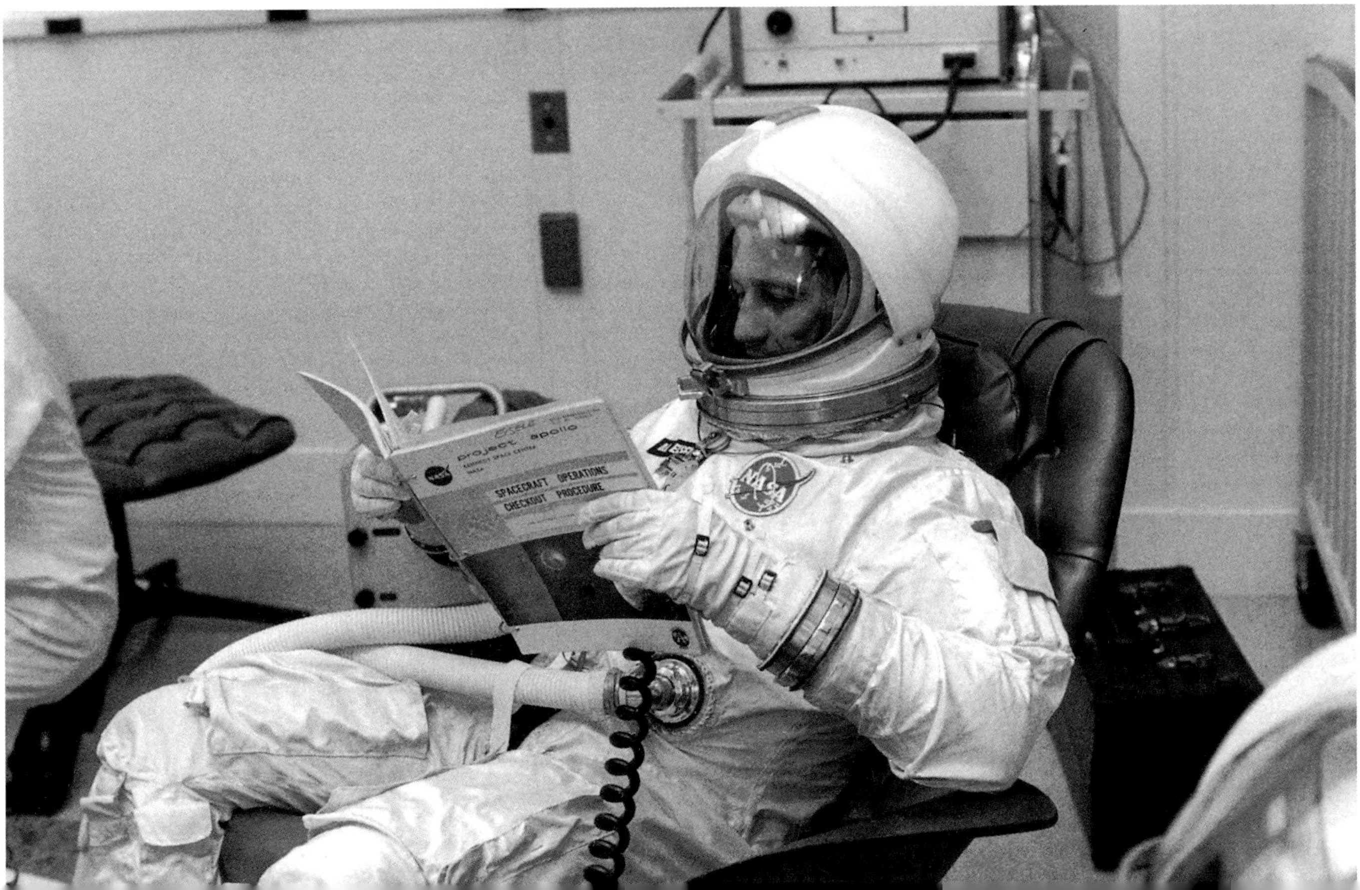

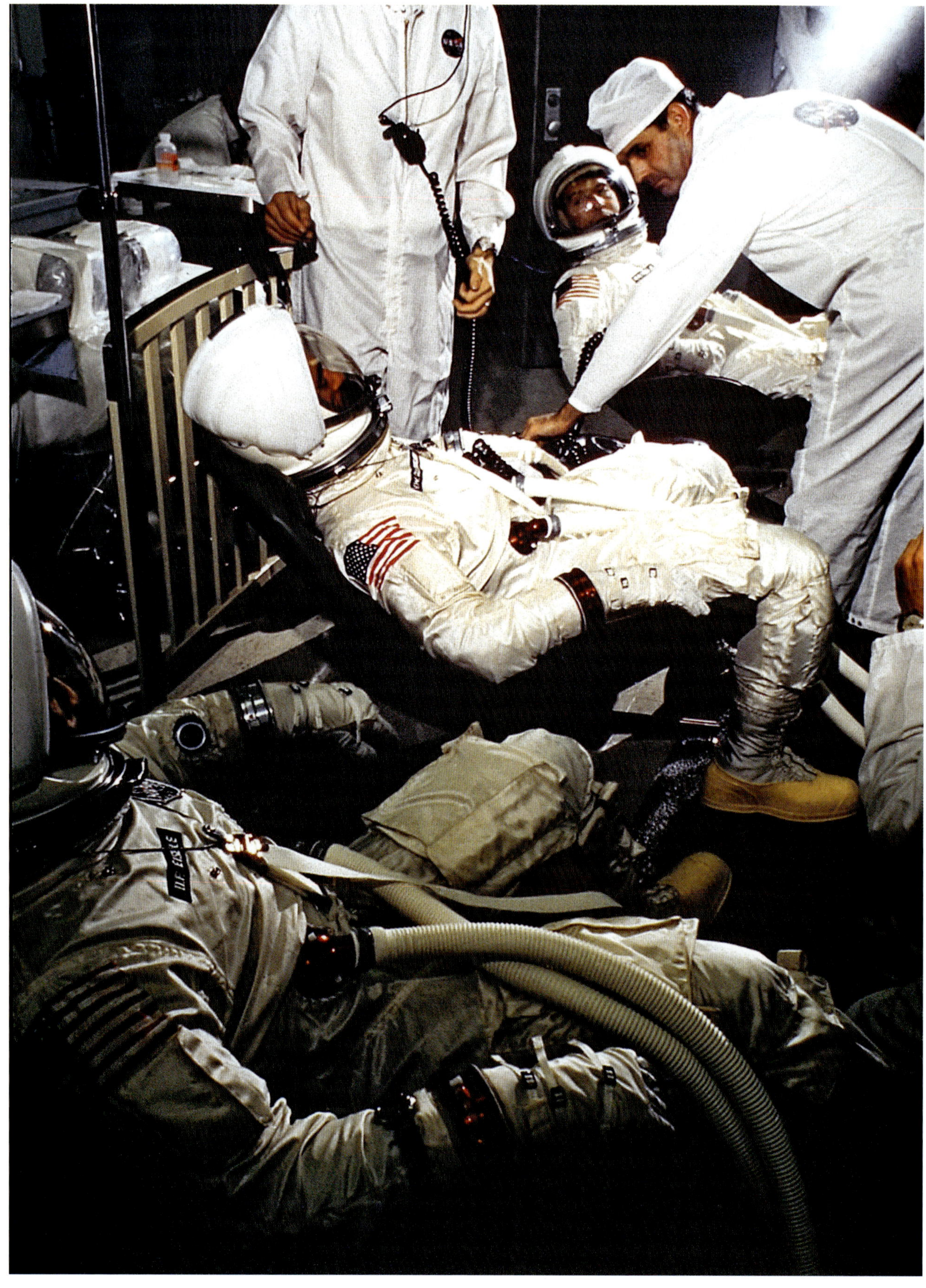

Technicians conduct final checks on the crew.
Top to bottom: Schirra, Cunningham, and Eisele.

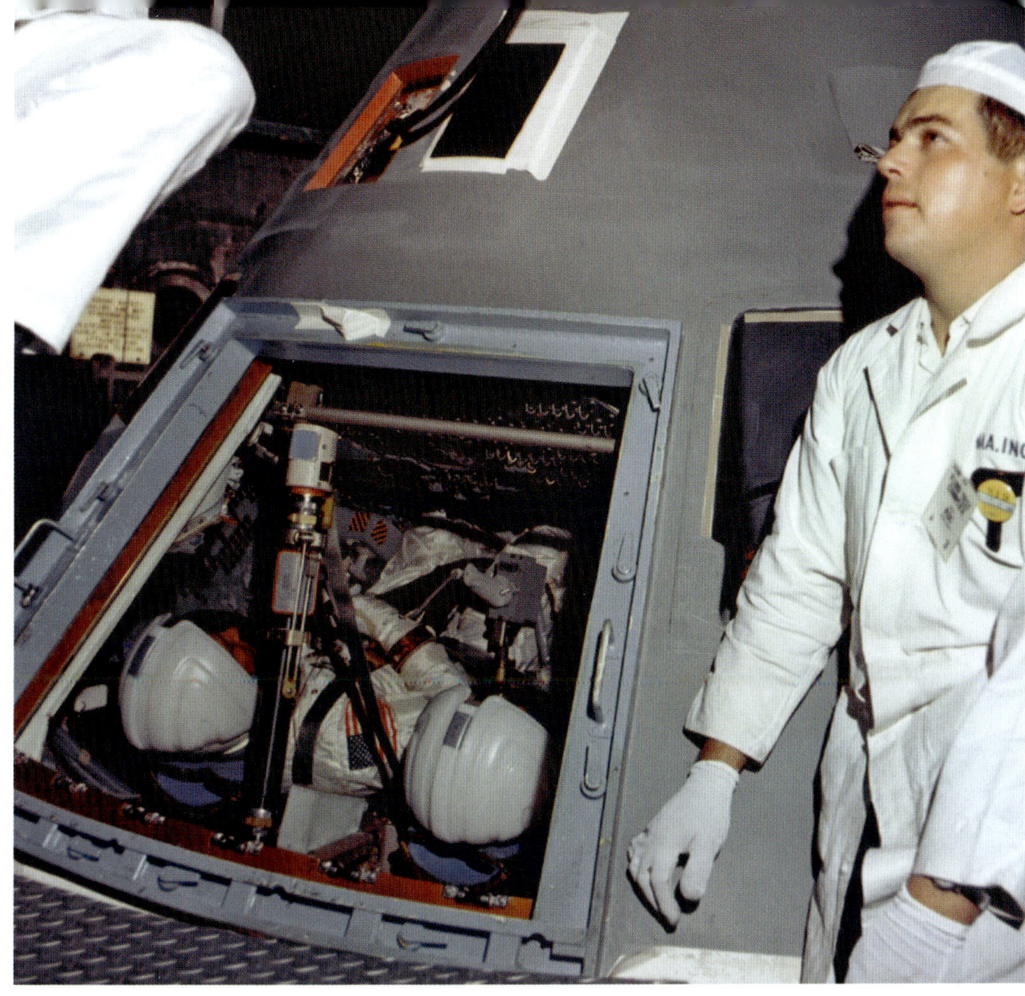

The astronauts get settled aboard Apollo I before the spacecraft is pressurized.

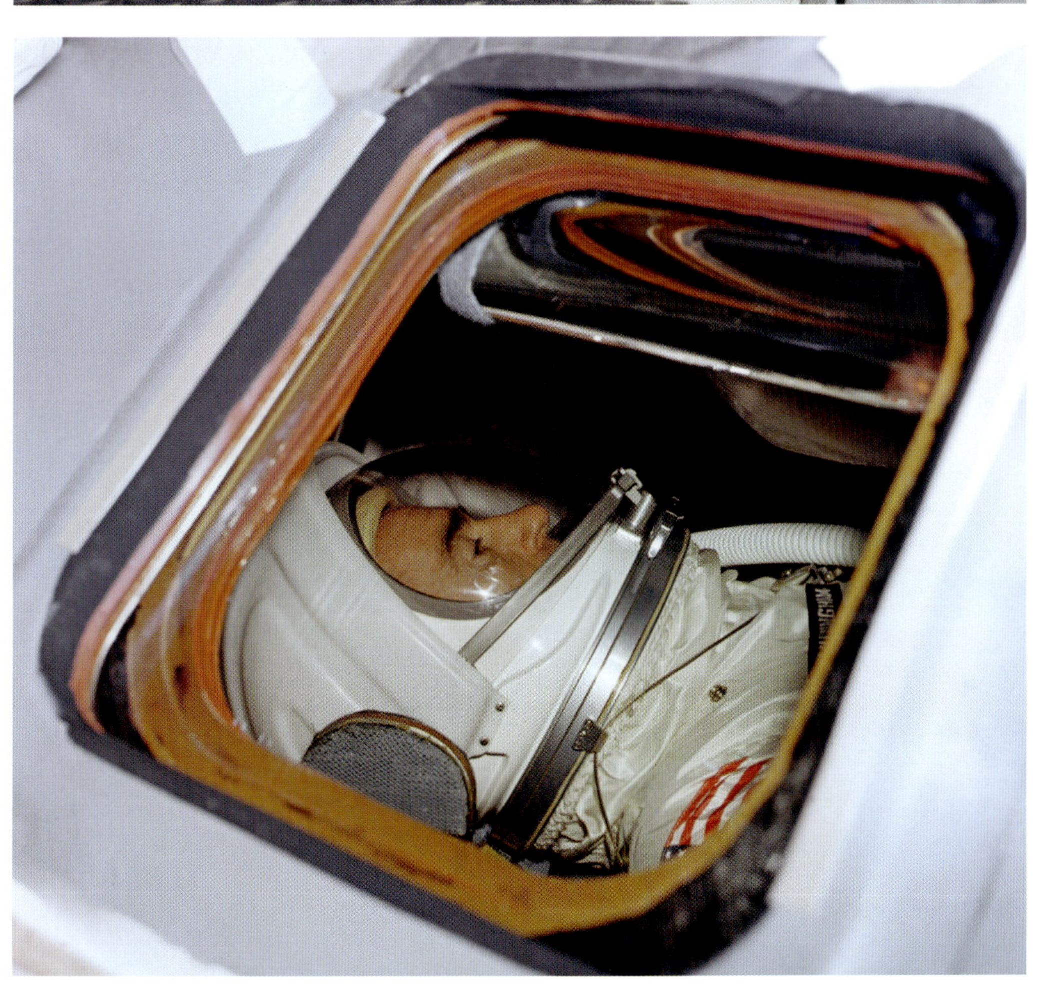

Cunningham is seen through side window no. 5 during the fifteen-hour test.

CHAPTER 8

January 3–10, 1967

With altitude chamber tests complete, two NAA technicians bolt the SPS engine nozzle to the thrust mount assembly in a workstand in the MSOB high bay on January 3. A gimbal ring allows the nozzle to move about 7 degrees up or down (pitch) or left or right (yaw), driven by a pair of motors. The circular opening on the aft heat shield at top is one of two propellant-fill-and-drain connectors.

A worker in a cherry picker bucket secures the CSM to the MSOB's bridge crane. The space between the CM and SM will be fully covered by a 26-inch-high faring.

The CSM is raised to next be mated to the SLA. The SM's radiators and RCS nozzles have protective covers.

The bridge crane moves the CSM toward Integrated Test Stand no. 1.

The crane begins to lower the CSM into the test stand.

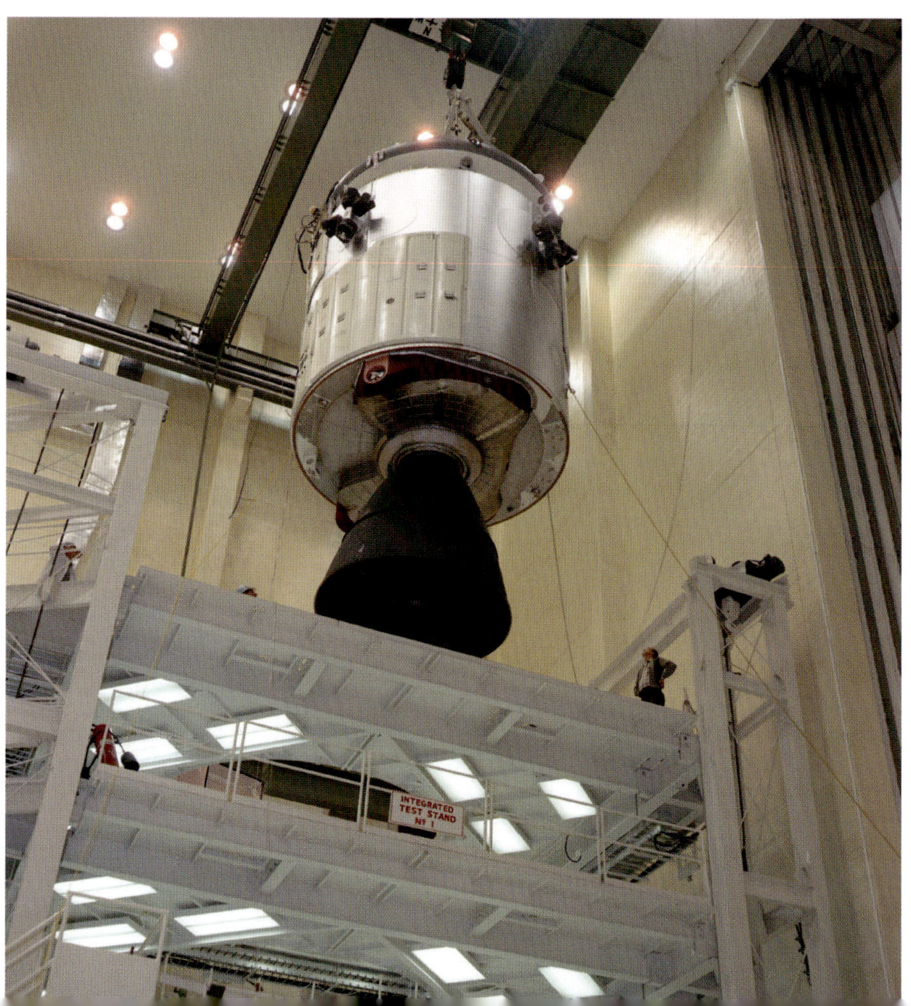

NAA and NASA technicians monitor operations as Apollo I is slowly lowered into the SLA just after midnight on January 4. Although the SM would not have a large S-band steerable antenna, first used on Apollo 8 for deep-space communications, its pivot point would be the bracket at upper left.

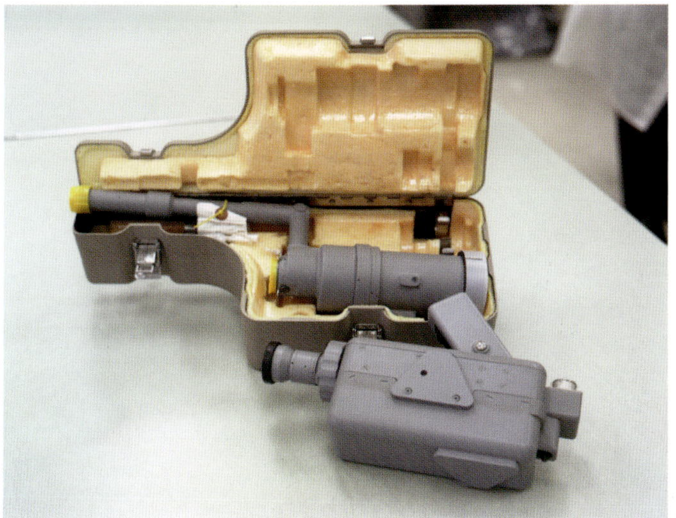

RCA's Astro Electronics Division developed this black-and-white slow-scan vidicon TV camera (with case) planned for Apollo I, shown at KSC on January 4. NASA managers envisioned using it primarily to monitor the crewmen during flight. The other item is an optional viewfinder.

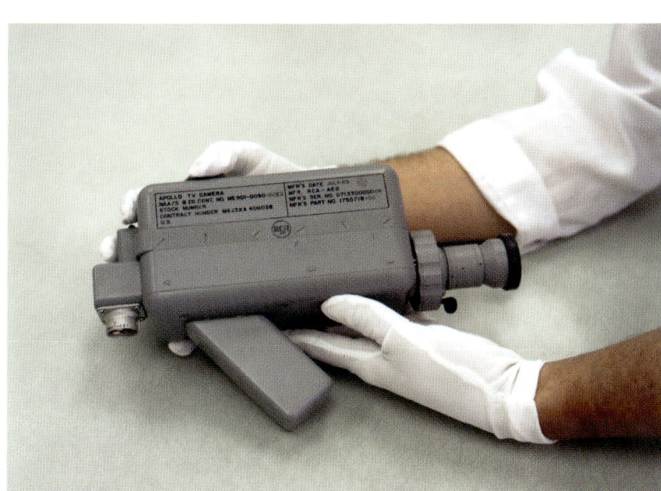

The opposite side of the camera shows its manufacturing information. The 4.5-pound camera provided only ten frames per second, with 320 lines of resolution. Only NASA's Corpus Christi, Texas, and Merritt Island, Florida, tracking stations could receive and convert the video signal for broadcast use, which required thirty frames per second.

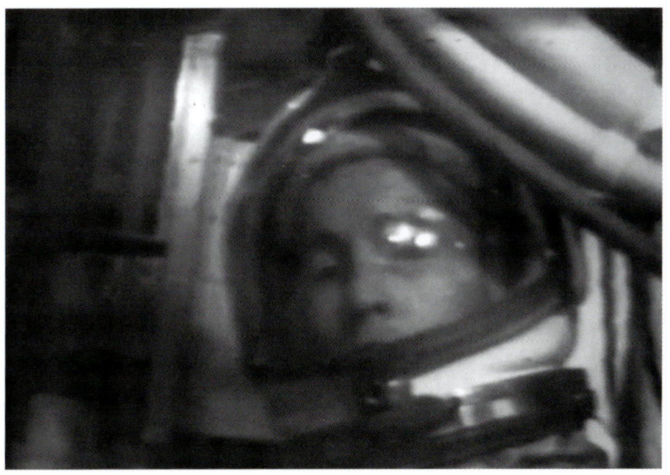

White appears in this screen shot during a test of the TV camera in the CM at Downey, on May 18, 1966.

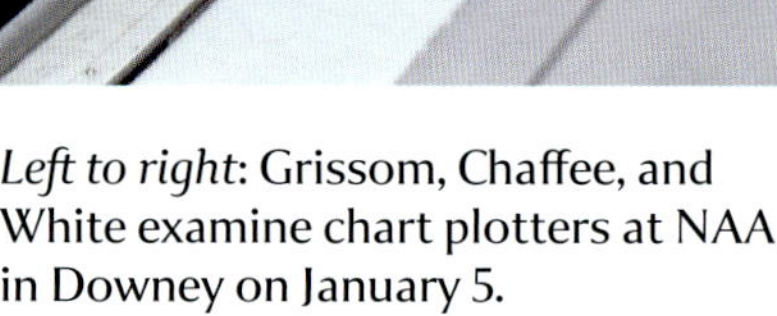

Left to right: Grissom, Chaffee, and White examine chart plotters at NAA in Downey on January 5.

White (*left*) and Chaffee pause next to the Block I CM Mission Evaluator at NAA on January 5. Frustrated by software problems with the AMS at KSC, the prime crew prefers to train here in the weeks before the mission. The Evaluator uses Apollo guidance and navigation software but could not interface with Mission Control.

Grissom (*second from left*), Chaffee, and White discuss training procedures with NAA Space and Information Systems Division managers, with the Evaluator in background.

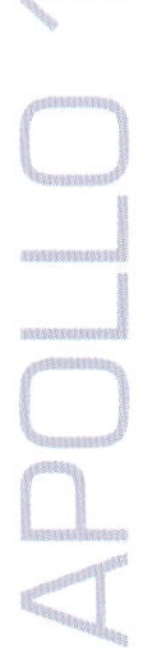

On January 6, the Apollo 1 CSM, stacked on its SLA, is moved from the MSOB to LC-34 for mating with its Saturn IB booster.

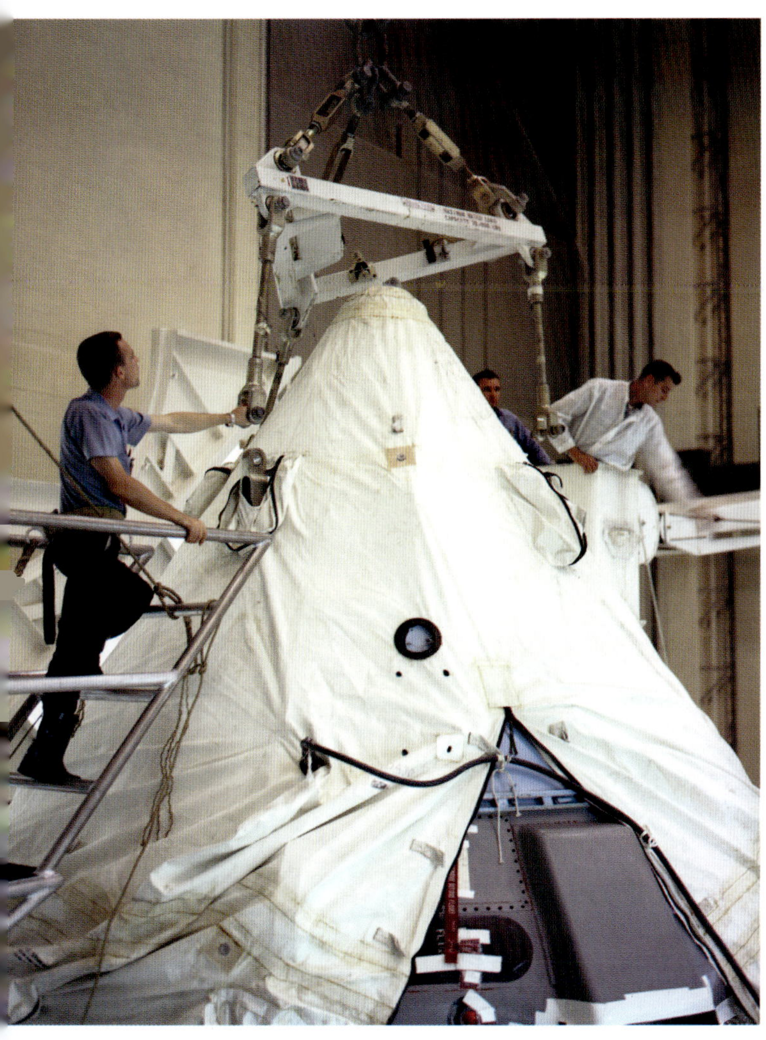

A worker on an extendable platform (*left*) and others on the cherry picker (*right*) attach the bridge crane to the CM, covered by a custom tarp, so it can be hoisted to be placed on its trailer.

NAA technicians gather at the base of the SLA in the MSOB. The spacecraft is in Integrated Test Stand no. 1, with its work platforms hinged upward before it is prepared for transport to LC-34.

Apollo 1 (*right*) waits at the door on its trailer. At left is CSM-017, being prepared for transport to the VAB. It would be flown on the unmanned AS-501 mission in November, the first Saturn V test flight.

The Apollo 1 spacecraft is ready to be towed through the 92-foot-high door of the MSOB's high bay about 8:00 a.m. EST for its move to the pad.

The 52-foot-tall stack passes the MSOB parking lot as it begins the almost hour-long trip from Merritt Island to LC-34, about 6 miles away.

Apollo I is on the move on its custom trailer.

The spacecraft heads southeast, crossing the drawbridge over the Banana River on the NASA Causeway East.

Apollo I leaves KSC property and enters Cape Kennedy AFS. Signage informs westbound drivers they are entering KSC, heading toward Merritt Island.

The Apollo I convoy moves through Cape Kennedy AFS.

The spacecraft moves northeast along Cape Road, passing the Cape Kennedy AFS Industrial Area.

The convoy approaches the pad perimeter at LC-34.

Apollo I arrives at the pad, where its Saturn IB booster, hidden by the service structure, has been in place since September. The white cherry picker at right will let workers attach the service structure's crane to hoist the spacecraft.

Technicians prepare to move the cherry picker into place near the top of the spacecraft to connect cables.

Technicians attach the hoist mechanism to the top of the Apollo I spacecraft.

The service structure crane has been attached.

Apollo I is hoisted off its transporter.

The spacecraft is slowly raised more than 140 feet to be mated with the booster.

The top of the Saturn IB's second stage is visible as the CSM is lifted.

Apollo I is raised into position. In the distance at left is the Vertical Integration Building; at right is the Solid Motor Assembly Building. Both support USAF Titan III and IV launches.

The spacecraft nears the top of the S-IVB second stage to be mated.

Apollo I is lowered toward its Saturn IB booster.

A technician works atop the white room as the service structure is moved in to surround Apollo 1. The tentative launch date is just more than six weeks away.

On January 10, a spare Rocketdyne H-1 engine (H-4062) arrives at the pad to replace H-4059, which was removed in December to replace its turbine wheel (the vendor had been approved only by the USAF). The engine was not reinstalled, however, because Rocketdyne had not explained a thrust anomaly during retesting in time for Apollo 1's planned launch.

Chrysler technicians raise the engine, still on its carrier, to a vertical position.

The technicians raise the engine vertically before its yellow carrier is removed.

The engine is hoisted up through the opening in the launch 'stand; during work at the pad, the opening is covered by a yellow platform, half open so the engine can be hoisted. The blockhouse is visible at lower right.

NASA technicians inspect the new H-I engine, one of eight in the Saturn's S-IB first stage, before it is moved into place with the others.

CHAPTER 9

January 17, 1967

Life magazine photographer Ralph Morse (*left*) shows (*left to right*) Slayton, White, and Grissom a gear-driven camera rig he devised for use in the Apollo Mission Simulator (AMS) in the Flight Crew Training Building at KSC on January 17. NASA managers had scheduled several hours that afternoon for news media photo shoots of the suited astronauts inside the simulator and at LC-34.

White examines a flashbulb as Grissom looks on.

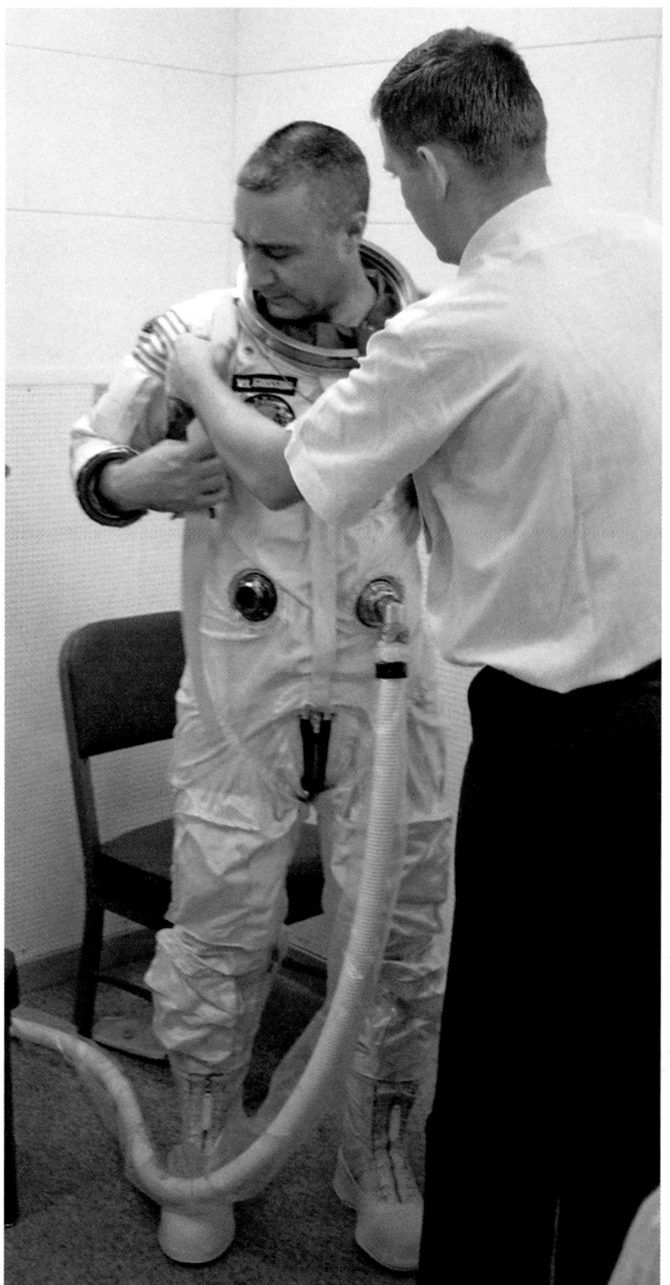

Morse shares a Polaroid test photo with White.

NASA suit technician Alan Rochford (*right*) suits up Grissom for the photography session near the simulator.

Chaffee's communications connector hangs from his neck ring as he gets suited. The crew patch has been moved from the shoulder to the front of the suit, although the mission is still not officially designated Apollo I.

Rochford (*left*) helps White with his water wings harness as Grissom looks on.

Chaffee and White get help with their harnesses; Grissom's is on. Technicolor's Larry Summers films the scene at right with an Arriflex 16 mm silent camera.

Rochford (*right*) makes an adjustment to Grissom's suit with the crewmen connected to ventilation equipment.

White adjusts his gloves.

Left to right: Chaffee, White, and Grissom head for the AMS, with Slayton behind them.

Top to bottom: Chaffee, White, and Grissom pose on the steps to the simulator.

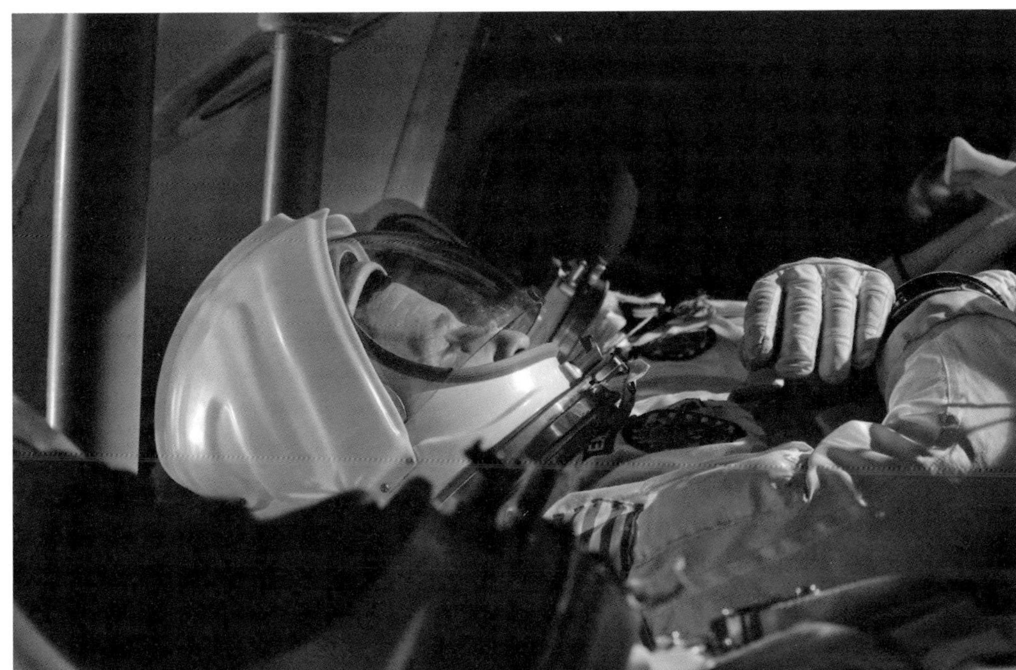

White's visor is partially closed.

Grissom is in the foreground in this side view.

White reaches below the main instrument panel.

Chaffee, White, and Grissom seen through side window no. 5.

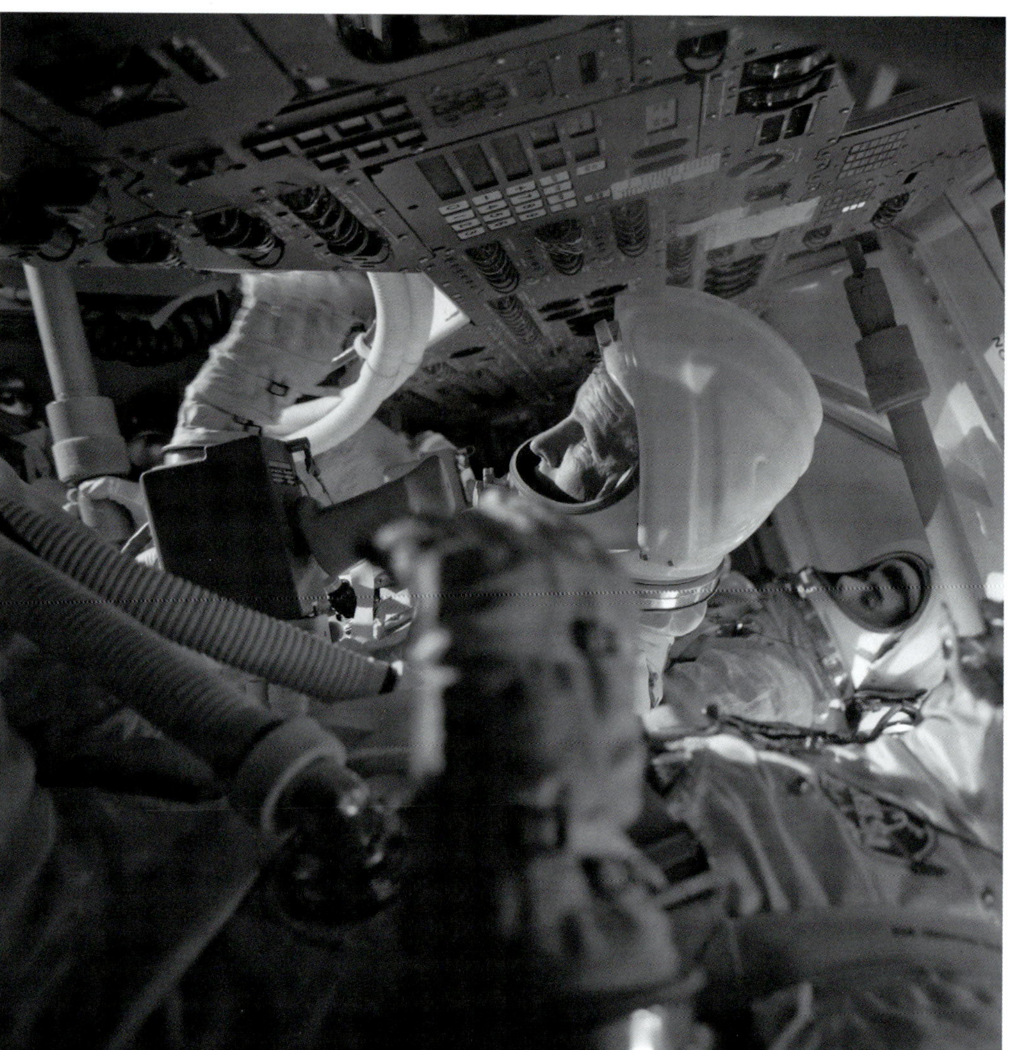

View through side window no. 1, showing Grissom, White, and Chaffee

White holds a television camera. RCA built six of the units.

A wide-angle view of the crew photographed with Ralph Morse's camera, seen in the photo on page 178. The photo captures both sides of the main instrument panel.

Chaffee tries out the television camera's viewfinder.

Grissom, White, and Chaffee (*not visible*) crouch in the lower equipment bay as a film cameraman rolls at bottom right. The simulator's two hand controllers are in the foreground.

Grissom and White exit the Flight Crew Training Building, heading to LC-34 to pose for news photographers.

Grissom and White board the Crew Transfer Van, a 1965 International Metro M-1200 van previously used for the final four Gemini launches.

Grissom, White, and Chaffee, with Rochford at left, head north along Cape Road to LC-34.

Chaffee and White arrive at the pad, with news photographers and reporters waiting at left.

Left to right: Grissom, White, and Chaffee pose with the LC-34 service structure surrounding their Apollo spacecraft and its Saturn IB booster. The structure was never rolled away to reveal the vehicle on the launch pedestal.

Grissom reacts to a comment as White and Chaffee look on.

Once considered irritable around reporters, Grissom was warming up, wrote AP space reporter Howard Benedict. Asked if he were mellowing, Grissom chuckled and called the photo sessions a necessary evil. “I guess it depends a lot on the amount of time we have. I remember back in Gemini, we had these big things going and I just didn’t have a day to give. We planned this one into our schedule this time . . . I recognize it has to be done.”

Grissom

White

Chaffee

Grissom, White, and Chaffee

Photographers jockey for good positions while photographing the crew. "At times there was bedlam," wrote Benedict (*left background*).

Grissom, White, and Chaffee. At the end of the photo session, Grissom surprises his two crewmates by pulling the cords to inflate their water wings. They return to the Flight Crew Training Building for more simulator time and later get press photos taken in their quarters at the MSOB. Six days later, the launch is officially set for February 21 at 10:00 a.m. EST, with a window until 3:30 p.m.

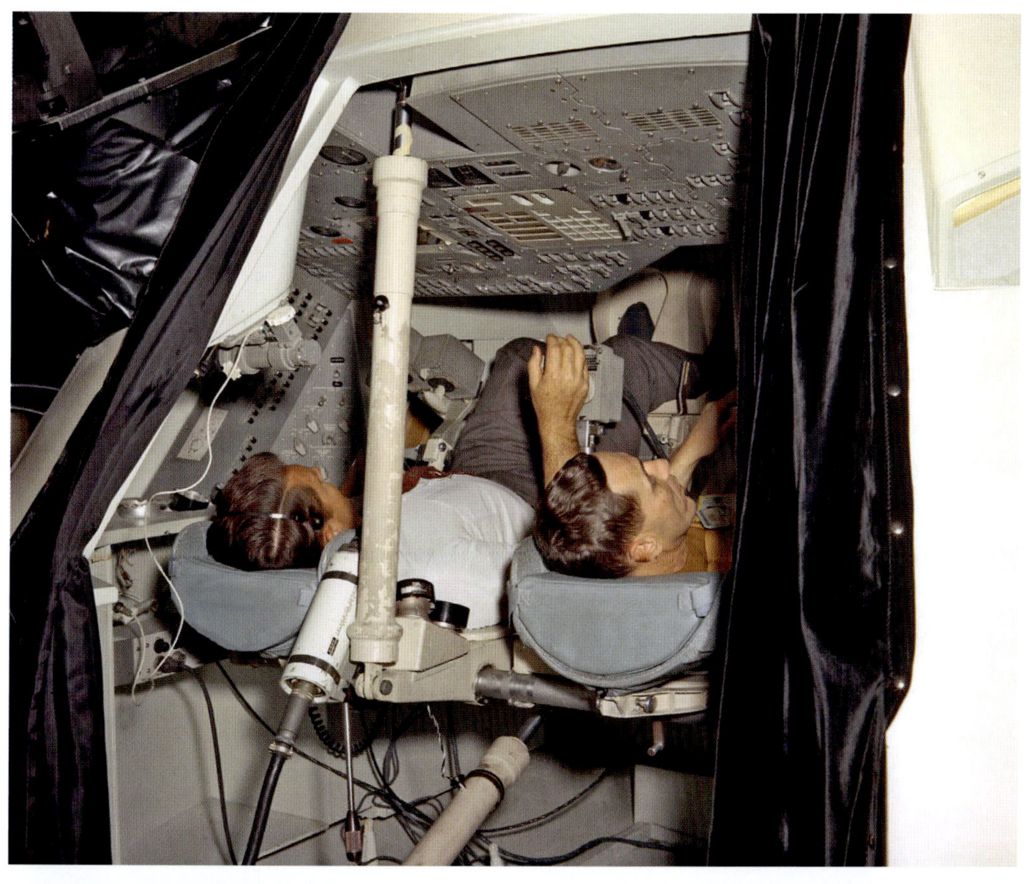

On the other side of the country that same day, backup crew members Schirra and Eisele train in the Block I Mission Evaluator at NAR in Downey.

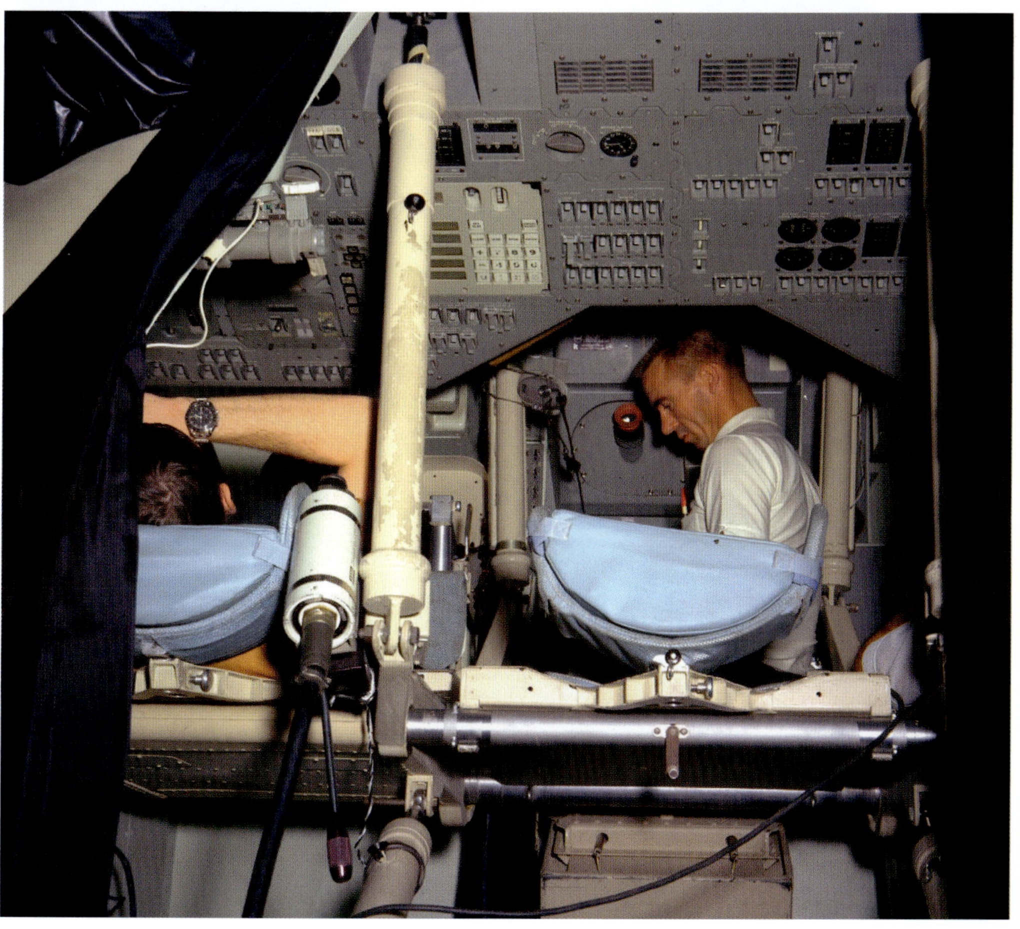

Eisele (*left*) and Cunningham in the Mission Evaluator

KSC director of launch operations Rocco Petrone is interviewed by NBC reporter Jay Barbree at LC-34 on January 18. Petrone contrasts scale models of the Saturn IB and a Gemini-Titan II.

CHAPTER 10

January 27–31, 1967

Grissom is greeted by DeArmond Mathews of KSC Security after the Crew Transfer Van arrives at LC-34 at 1:00 p.m. EST on Friday, January 27, for an abbreviated rehearsal countdown.

Grissom steps from the van. The astronauts have been breathing pure oxygen to purge their bodies of nitrogen to prevent the bends when the CM's cabin pressure drops during ascent. *Bill Taub archive*

Chaffee follows Grissom from the van, with Matthews at left. *Bill Taub archive*

A fish-eye lens captures Grissom walking underneath an overhang as Chaffee exits the van. They will be the first to ingress the CM.

Chaffee, carrying his ventilator, passes ground personnel. *Bill Taub archive*

Grissom and Chaffee head toward the umbilical tower elevator, with Rochford behind them.

Grissom approaches the elevator, which could reach the ground in thirty seconds and is the only means of astronaut emergency escape.

Grissom and Chaffee enter the elevator, with Rochford carrying backup ventilators.

Rochford sets down the ventilators, with Chaffee and Grissom behind him. They'll head to the 224-foot (spacecraft) level, then Rochford will return to pick up White.

White (*right*) follows Deke Slayton from the van.

White heads toward the elevator, followed by suit tech Joe Rebokus (*left*) and Slayton (*right*).

★ Incident Timeline: January 27–28 (all times EST)

6:00 a.m.	test begins
1:00 p.m.	crew arrives
2:45 p.m.	hatches closed
6:31 p.m.	fire breaks out
6:36 p.m.	hatches opened
6:40 p.m.	firefighters arrive
6:45 p.m.	fire extinguished, three physicians arrive
1:55 a.m.	bodies removed

Grissom (*right*) leads Chaffee across the access arm to the CM.

Slayton, White, and Rochford prepare to ascend to the 224-foot level. The crew ingresses by 1:19 p.m. The fatal fire breaks out at 6:31 p.m., with the countdown in a hold at T-minus ten minutes.

The service structure is silently illuminated on the evening of January 28. *AP photo*

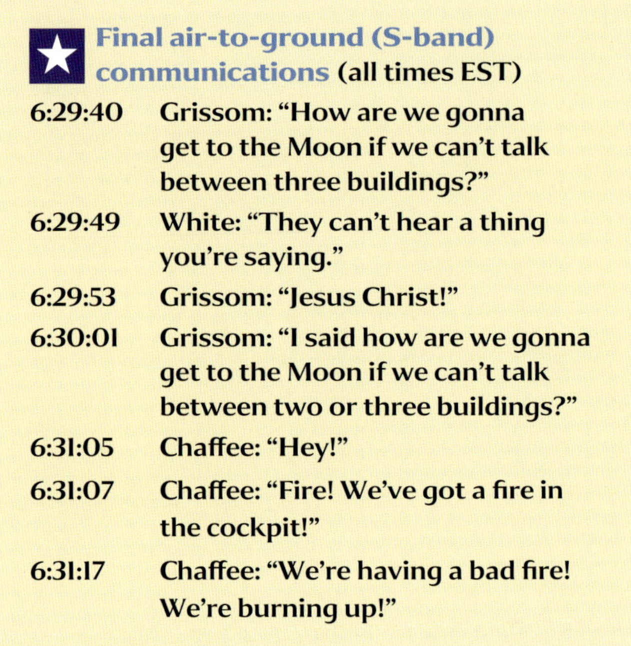

Final air-to-ground (S-band) communications (all times EST)

6:29:40	**Grissom: "How are we gonna get to the Moon if we can't talk between three buildings?"**
6:29:49	**White: "They can't hear a thing you're saying."**
6:29:53	**Grissom: "Jesus Christ!"**
6:30:01	**Grissom: "I said how are we gonna get to the Moon if we can't talk between two or three buildings?"**
6:31:05	**Chaffee: "Hey!"**
6:31:07	**Chaffee: "Fire! We've got a fire in the cockpit!"**
6:31:17	**Chaffee: "We're having a bad fire! We're burning up!"**

Areas of the CM's heat shield were darkened by the fire, as seen the next day on the Level 8 platform of the service structure, as a Cape police officer stands watch. A section of the white boost protective cover had been removed for the test. A red fire extinguisher cart is at right. Twenty-seven pad workers were overcome by smoke trying to rescue the astronauts, and two were hospitalized overnight. These white room photos were taken by NASA senior photographer Bill Taub.

The opposite side of the CM shows the most-extensive damage, with flames apparently erupting from the helium pressurization subsystem access panel (*left*) and the oxidizer control panel for the CM's RCS (*right*). The white room is in place against the hatch at left.

Close-up of the oxidizer control panel. The CM's outer shell honeycomb structure is visible from the heat.

In the small white room, Dale Carothers of the Spacecraft Operations Directorate inspects the interior of the CM, showing extensive charring from the fire in the pressurized pure-oxygen environment. Access from Level 8 was through one of two side doors.

Suit hoses had been cut for removal of the bodies, still in their suits. White's center couch is covered with debris, including pages from the flight plan; Chaffee's seat pan is dropped at right, indicating he tried to leave the spacecraft.

The pad leader's desk and work area on the north side of Level 8 (Level 7 below surrounds the SM). The lead pad technician's desk is in the background at right. All the fire extinguishers are later found to be past their inspection dates.

Lower left: A singed notebook on the pad leader's desk

Below: Debris and charred materials cover the desk. A spacecraft flow chart and CSM diagram are taped to the railing.

Above: Three NAA engineers served as pad leaders: Lewis Curatolo (first shift), Don Babbitt (second shift), and John Murphy (third shift). Babbitt helped remove the BPC hatch in near-zero visibility. Curatolo relieved him because of smoke inhalation, and colleagues initially feared that Babbitt had died. Murphy relieved Curatolo at 8:00 p.m.

Top left: The phone and communications box at the pad leader's desk. Reluctant to report the fatalities on the communications circuit, Babbitt initially advised the ground he wouldn't describe what he'd seen.

Left: The CM inner hatch leans against a file cabinet on Level 8.

Two Cape police officers keep watch at LC-34 on January 28. Additional officers are assigned to the complex, which had immediately been impounded.

George Alexander, Florida bureau chief for *Aviation Week & Space Technology*, kneels to take notes on Level 8 on January 29. He had been unanimously selected by colleagues to be the "pool" reporter allowed to view the spacecraft, then share his findings with them. The CM's BPC hatch is next to him, with the inner hatch at right. The pool photographer stands at left.

Alexander examines the CM's interior. "It was the most appalling sight you can imagine," he said. The exterior of the white room is behind the spacecraft. *AP photo*

On the morning of January 30, Grissom's casket leaves the Bioastronautics Operational Support Unit (BOSU), where the crewmens' bodies had been taken for autopsies early on January 28.

Six airmen and two sailors approach the hearse; LC-37 is in the distance.

Grissom's casket, in a protective wooden shipping container, is aboard the hearse.

The hearse bearing Grissom's casket departs the BOSU, heading toward Hangar Road, with those carrying White and Chaffee following. The slow procession, which takes about ten minutes, then turns left onto Skid Strip Road for about another mile to the runway.

The hearses pass mourning government and contractor employees along Skid Strip Road.

The hearses arrive at the Skid Strip, the Cape's 10,000-foot runway.

An NBC pool TV camera provides live coverage to NBC and CBS. The pool uses equipment from WFGA-TV, the NBC affiliate in Jacksonville, Florida. The black-and-white RCA TK-31 camera uses a Rank-Taylor-Hobson zoom lens.

Members of the Apollo 204 Review Board, established two days earlier, arrive for the ceremony. *At left*, Patrick AFB commander Col. Joseph Williams walks with Langley Research Center director Floyd Thompson, board chairman. Behind them is John Williams, with Col. Charles Strang to his right. E. Barton Geer is at center, with Charles Mathews, Apollo Applications Program director and a board consultant, at right. About three hundred people attend the event; the review board later reconvened at the Cape.

The three hearses are parked for the transfer, with LC-36 in the distance. A five-man color guard and an honor cordon of USAF and Navy enlisted men take part as a USAF band plays.

White's casket joins Grissom's on a Boeing C-135B Stratolifter as "The Air Force Hymn" plays.

Chaffee's casket is loaded to "Faith of Our Fathers." The aircraft departs at 10:23 a.m. EST for the flight to Andrews AFB, Maryland.

After a brief noontime ceremony at Andrews AFB, sailors and airmen prepare to load Grissom's and Chafee's caskets into hearses for the trip to Arlington Cemetery, Virginia, where they will remain overnight. An honor guard representing all branches of the armed services salutes in the background. The aircraft then continues to Stewart AFB, New York, with White's casket.

The US flag flies at half-staff at MSC Headquarters on January 31; memorial services had been held in Houston during the previous two days. Motels in Cocoa Beach, Florida, also lower their flags.

A horse-drawn caisson bearing Grissom's casket prepares to move to his gravesite in Section 5 at Arlington on January 31. Escorts in the foreground (*left to right*) are astronauts John Glenn, Gordon Cooper, and John Young; Mercury flight surgeon Dr. William Douglas; and Mercury program official Walter Williams. At left are astronaut Scott Carpenter and MSC director Robert Gilruth.

The USAF honor guard prepares to fold the flag from Grissom's casket about 9:00 a.m. as the remaining six Mercury astronauts stand at attention. Seated in the first row are (*left to right*) Grissom's widow, Betty; their sons, Mark, thirteen, and Scott, sixteen; his parents, Mr. and Mrs. Dennis Grissom; and President Lyndon Johnson. This ceremony, as were the other two that day, is accompanied by three volleys from a firing squad and a three-jet flyover in the missing-man formation.

President Johnson presents the flag to Betty Grissom, with Mark and Scott at right, as Schirra looks on. Louise Shepard is in the brown coat behind them.

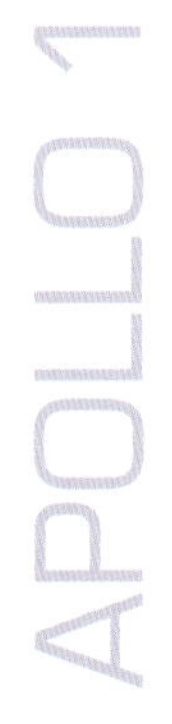

At 1:00 p.m. that afternoon, astronauts Mike Collins and Dave Scott (*right*) escort Chaffee's casket at Arlington. Walt Cunningham is on the opposite side. Astronauts Dick Gordon, Alan Bean, and Donn Eisele are hidden by the caisson.

Chaffee's casket is moved by a USN honor guard to the gravesite next to Grissom's in Section 5.

Left to right: Chaffee's widow, Martha; their daughter, Sheryl Lyn, nine; their son, Stephen, five; his parents, Mr. and Mrs. Donald Chaffee; and President Johnson at the gravesite.

White lies in state in the Chapel at the US Military Academy at West Point, New York, from 10:00 to 10:30 a.m. on January 31. The service begins with a cadet chorus singing his alma mater.

An Army honor guard stands watch near White's casket. The first lady, Lady Bird Johnson, and Vice President Hubert Humphrey represent the president at the services.

White's widow, Patricia, is flanked by his brother, James, and his fellow Gemini 4 astronaut Jim McDivitt following the burial at West Point Cemetery at 11:00 a.m.

CHAPTER 11

January 31–July 1967

Irving Pinkel, chief of the Fluid Systems Components Division at Lewis Research Center (*center*), hands a bagged logbook from the Apollo 1 CM to Homer Carhart (*left*), head of the Fuels Branch at the Naval Research Lab, at LC-34 on the evening of January 31. Thomas Horeff, manager of the FAA's Propulsion R&D Program, is at right. The three are members of the Origin and Propagation of Fire Panel, one of twenty-one task panels established by the Apollo 204 Review Board.

Horeff, Pinkel, and Carhart examine the CM's RCS oxidizer control panel on Level 8 of the service structure. The combustion experts and other investigators spend more than two hours inspecting the spacecraft.

The LES is removed and lowered by a hoist on February 1. The Lockheed Propulsion Co., northeast of Redlands, California, provides both the launch escape and the pitch control motors of the LES, which uses a solid polysulfide fuel. The Thiokol Chemical Corp. of Ogden, Utah, provides the tower jettison motor.

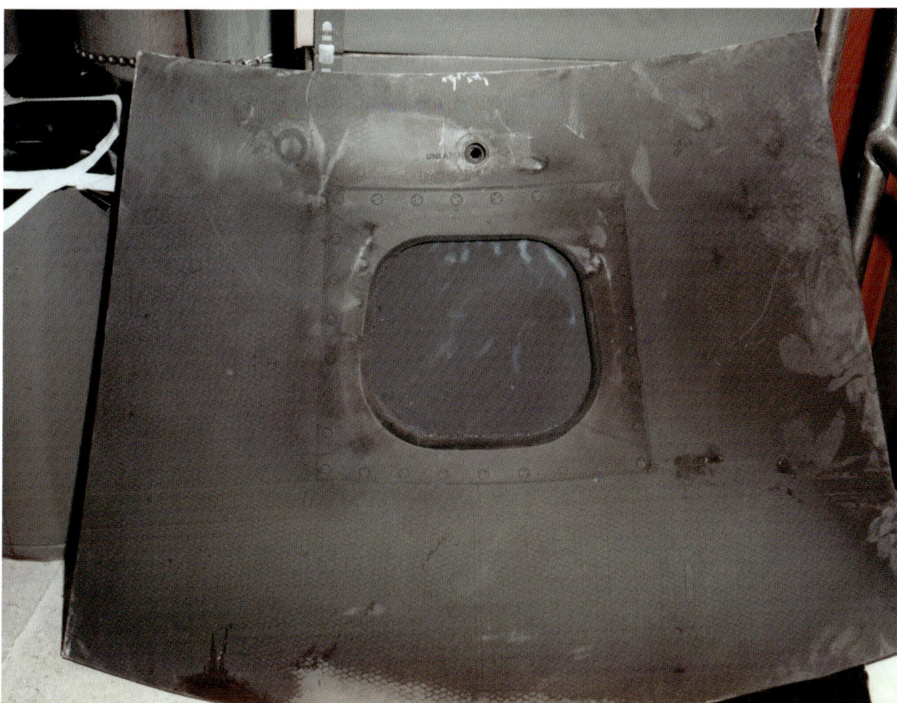

The CM's outer hatch would be sent to the Pyrotechnic Installation Building (PIB) in the KSC Industrial Area on February 3, as would the entire CM two weeks later.

The inside of the inner hatch, seen after arrival in the PIB on February 3, shows the latching system.

The eight Apollo 204 Review Board members and their counsel meet in the MSOB auditorium on the morning of February 4 to receive a report from the Disassembly Panel. *Left to right*: astronaut Frank Borman; John Williams, KSC spacecraft operations director; George Malley, counsel (Langley); Floyd Thompson, chairman; E. Barton Greer, associate chief, Flight Vehicles & Systems Division, Langley; George White, director, Apollo Reliability and Quality, Apollo Program Office, NASA Headquarters; Col. Charles Strang, chief of the USAF Missile and Space Safety Division; and Max Faget, MSC director of engineering. The board meets almost daily from January 28 through March.

The next day, investigators visit the AMS in the Flight Crew Training Building, where astronauts run through emergency provisions for an in-flight fire, which basically consists of donning pressure suits and depressurizing the cabin. Borman, who'd been the first astronaut to enter the spacecraft after the fire to verify switch positions, is flanked by Col. Strang (*at left*) and Thompson (*at right*).

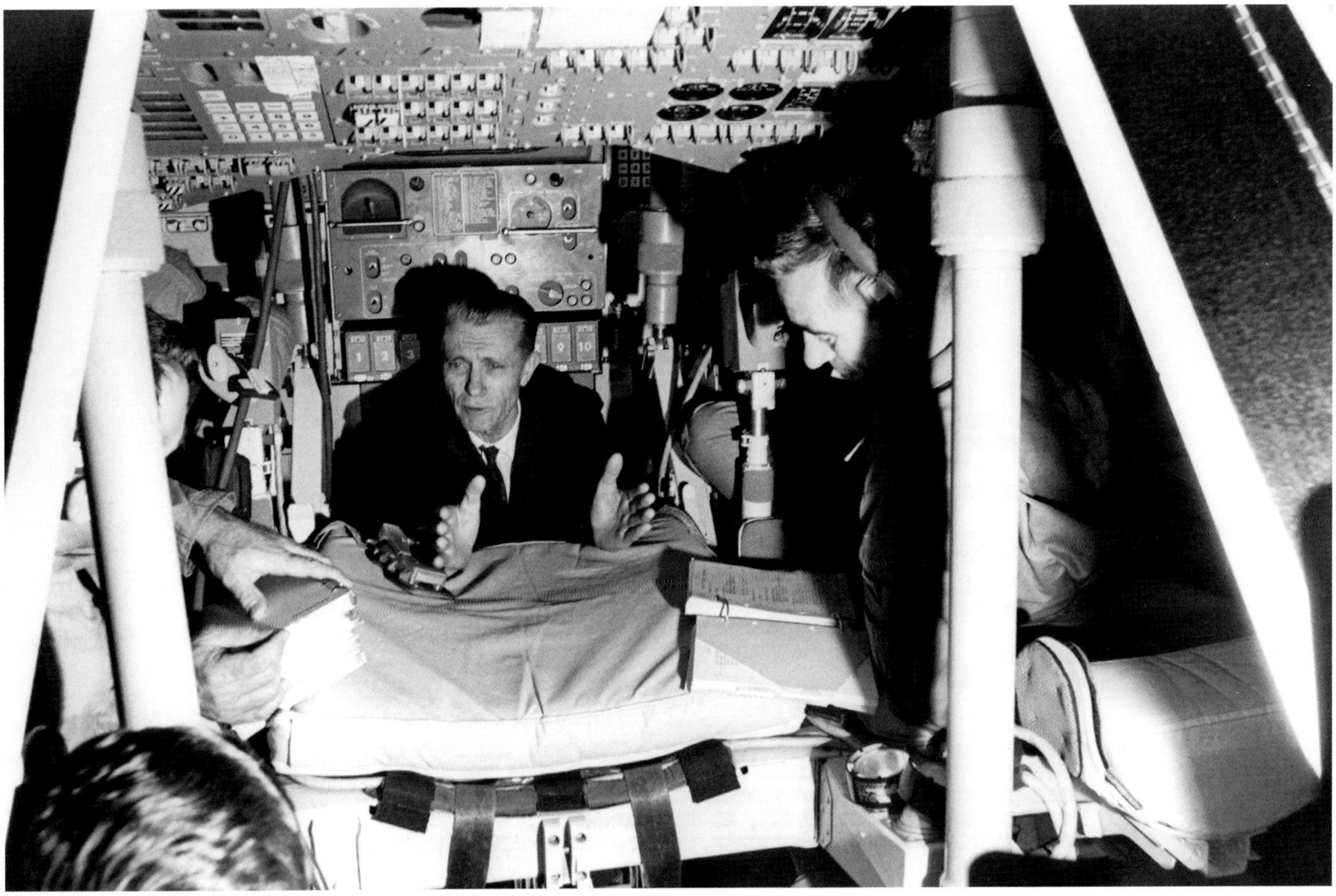

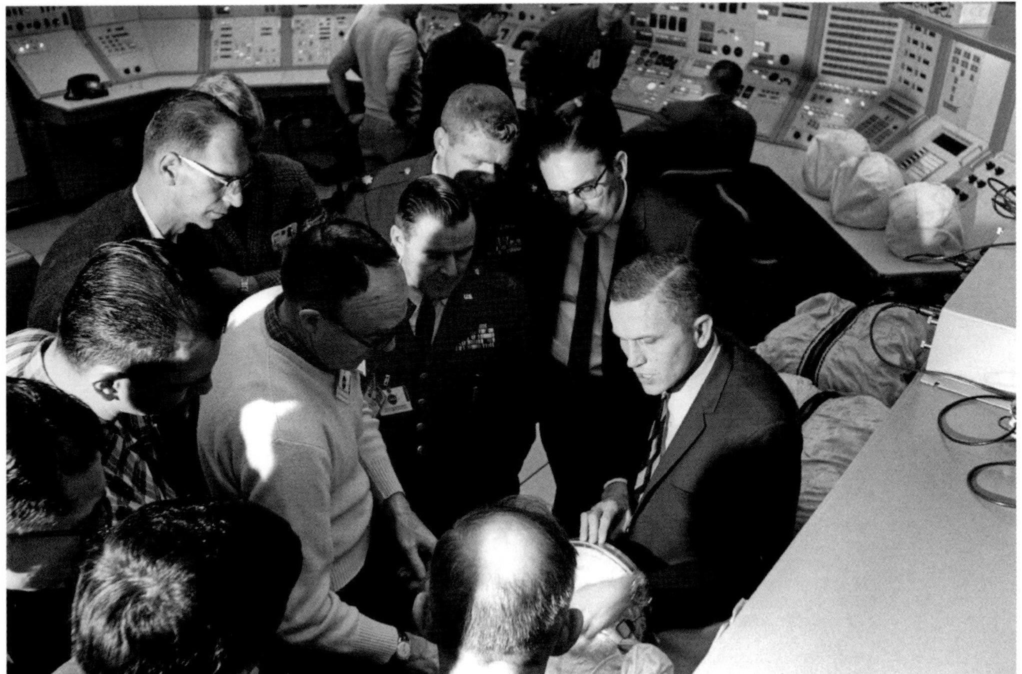

Thompson is briefed by astronauts inside the simulator, including Walt Cunningham (*right*), Jim Lovell (chairman of the board's panel on in-flight fire emergency procedures), and Donn Eisele.

Borman shows investigators an AIC helmet. Wearing glasses (*at left*) is KSC's Bob Moser, a board consultant and former Mercury test conductor.

Borman demonstrates the helmet's visor, worn by astronaut Ron Evans, who is also on Lovell's panel. Col. Charles Strang is at center, with Carhart at far right.

At LC-34, extensive photo documentation of the CM's cabin is underway by NASA's Bill Taub, using incandescent spotlights to avoid fluorescent color shifting. Grissom's side of the main instrument panel shows damage, with the screen for the attitude indicator at center.

The Display Keyboard, used by the crew for computer inputs, is covered with soot.

Grissom's headrest and couch, with a suspension strut in the foreground

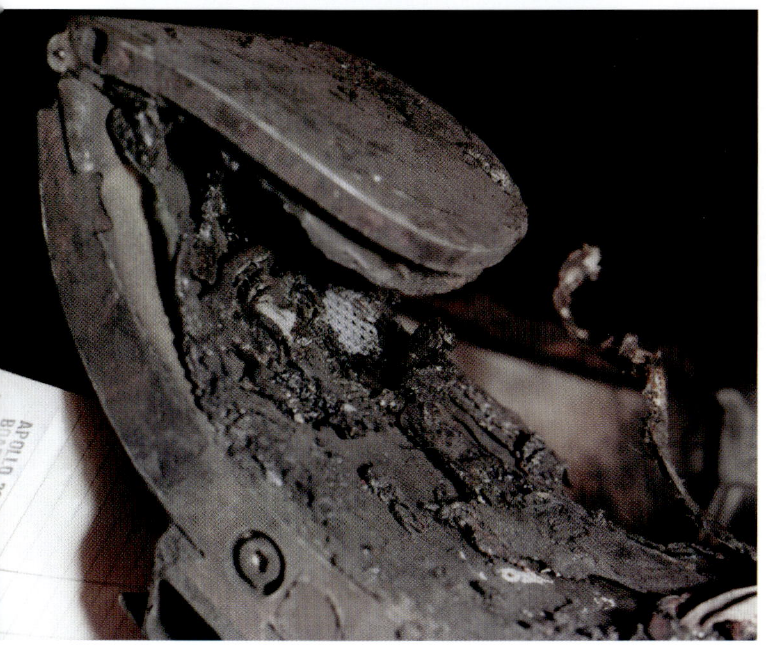

The left side of White's headrest from his center couch

White's Omega Speedmaster wristwatch is next to his headrest. All debris from each couch is carefully photographed and bagged.

A shelf area under a side window shows an unaffected area from soot after a book was removed.

This corner of the CM is above the right-hand lower equipment bay on Chaffee's side of the spacecraft.

Rendezvous window no. 2 on Grissom's side of the CM

The gas chromatograph compartment in the left side of the lower equipment bay, near Grissom's feet, is near the area where the fire likely began. Two 16 mm sequence cameras and a camera power cable were stowed loose on the floor.

The CM floor beneath White's feet shows stowed helmet covers and a lithium hydroxide canister storage box.

The ECU in the lower left hand equipment bay (Grissom's side), with side window no. 1 at top left

This area near the floor in the lower forward section of the left equipment bay below the ECU was cited by the board as where the fire most likely started, on the basis of studies of burn patterns. The report also said the most probable cause was an electric arc in a power cable in this area (a section of the cable is in the center foreground).

Grissom's side of the spacecraft. Side window no. 1 is at right center, and rendezvous window no. 2 is at center. The hatch is on the left, and Grissom's couch is at lower right.

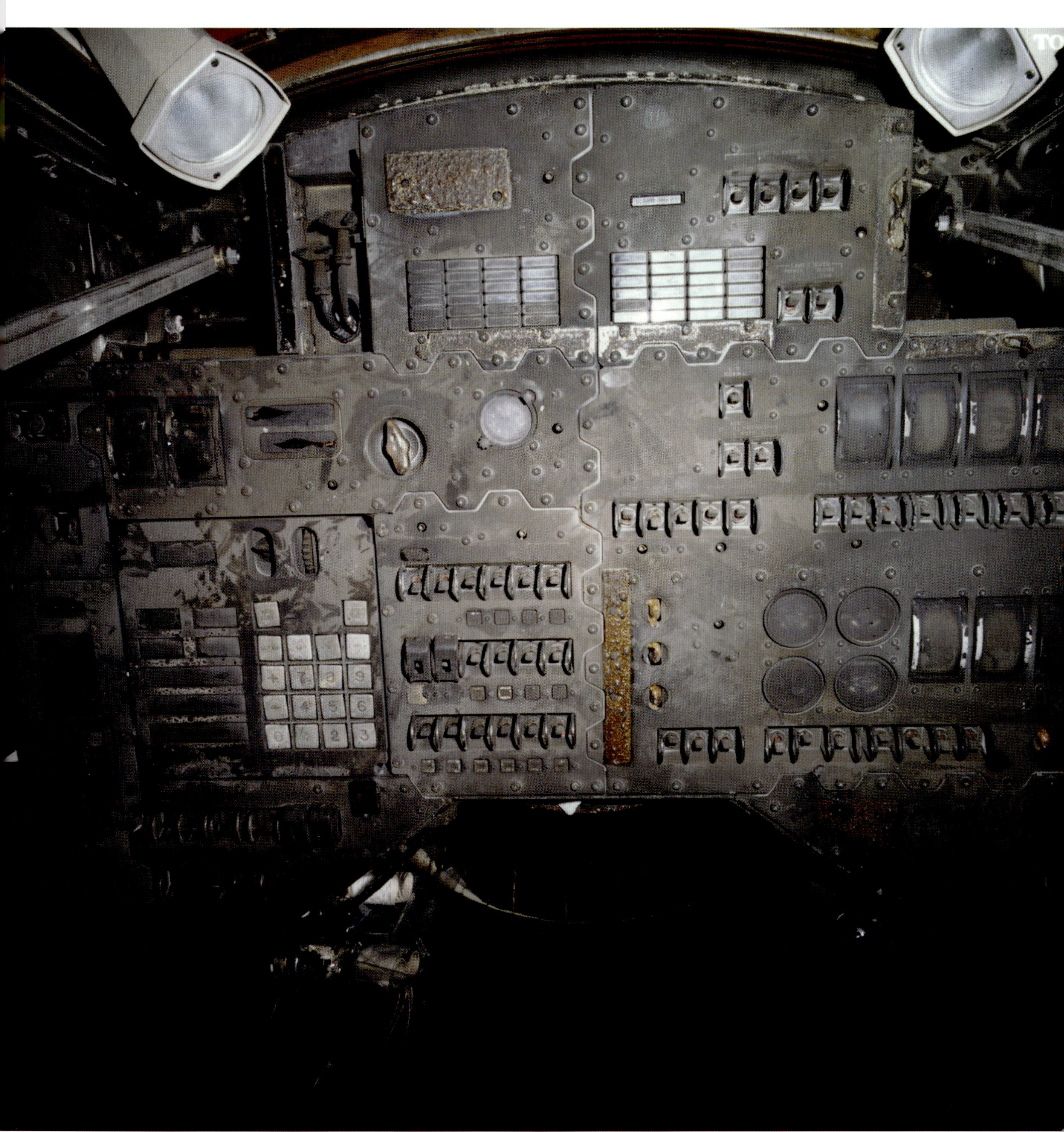

White's view of the main instrument panel from his center couch. Much of the cabin interior is covered with soot.

The right section of the lower equipment bay is where Chaffee's feet would have been.

The review board quickly established detailed procedures for removing items from the CM at the pad. First, a panoramic photo and a series of stereo photos were taken to document the CM's as-found condition. A piece of plywood across the couches was used for initial access. After the couches were removed, a false floor of transparent squares was suspended from the couch strut attach points to provide access without disturbing evidence. Two persons at a time were permitted inside the CM to expedite component removal, which continued on a 24/7 basis. Each item was inspected, photographed, tagged, and sealed in a plastic container for transport under security to the PIB (about five thousand photos were eventually taken). On February 16, the board decided that the CM could be demated and moved to the PIB, which provided better working conditions and space for full disassembly.

This access panel was used to fuel the yaw engines of the CM's RCS.

A false floor with transparent squares is installed in the CM on February 5 to provide access for removing components at the pad. Each item is then bagged for shipment to the PIB, where the components and spacecraft would be examined.

On February 17, a bridge crane lifts the wrapped Apollo 1 CM from its SM during demating operations at LC-34 after the board agreed it could be moved.

The 9,400-pound CM is lowered that afternoon. The cable run to the blockhouse extends out of frame at upper right; rails for the service structure are at right.

Blackened areas on the heat shield are visible as the CM is brought down. Two slings are in place in case the aft heat shield had been compromised and could separate.

Bendix technicians watch as the CM is lowered toward a flatbed trailer with a blue handling ring. Wrapped cables and lines from the CM/SM umbilical hang down at center.

Bendix workers secure the CM for the 5-mile trip to the PIB at KSC.

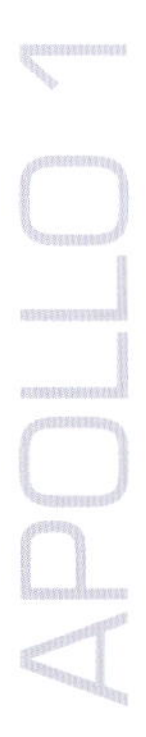

The Apollo I CM is raised from its trailer inside the PIB on the evening of February 17. The wrapping has been lifted on the hatch side. The facility was originally the Weight and Balance Building, used to install pyrotechnics for Gemini spacecraft. The review board's twenty-one subcommittees, meanwhile, are preparing to submit their interim reports.

The combination is lowered onto a trailer by Bendix workers.

The SM and SLA begin the trip to the MSOB on February 23. Senior NASA officials return to Washington, DC, that day after two days of briefings at KSC by Apollo managers on how the program's schedule would be affected. The SM was eventually scrapped at North American Rockwell (NAA's successor) in Downey.

On February 23, a bridge crane slowly lowers Apollo I's SM and SLA after demating, with the IU exposed at bottom.

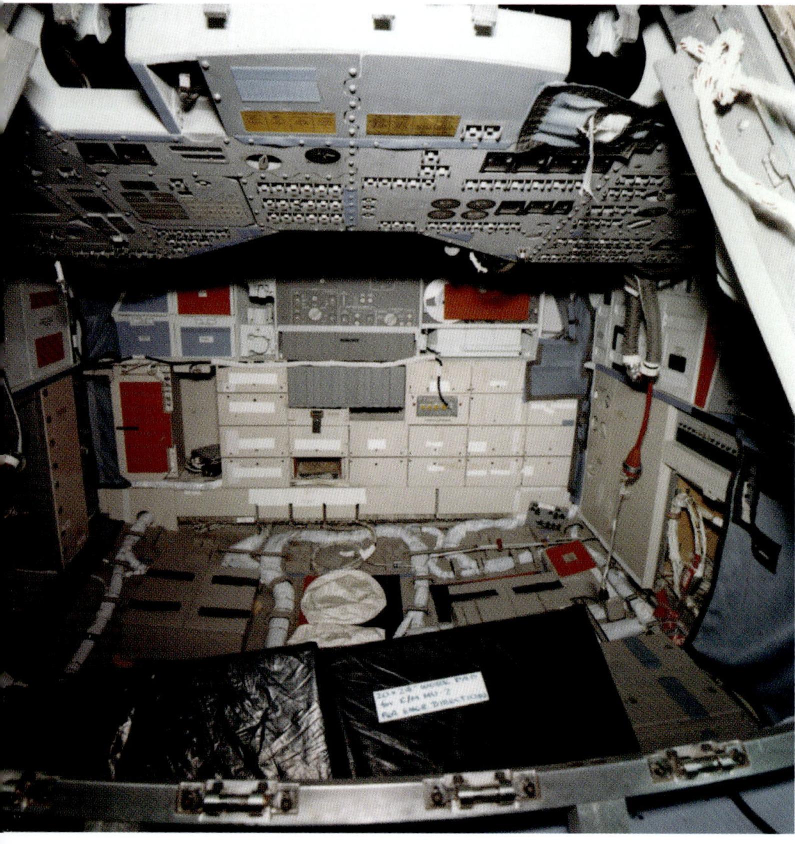

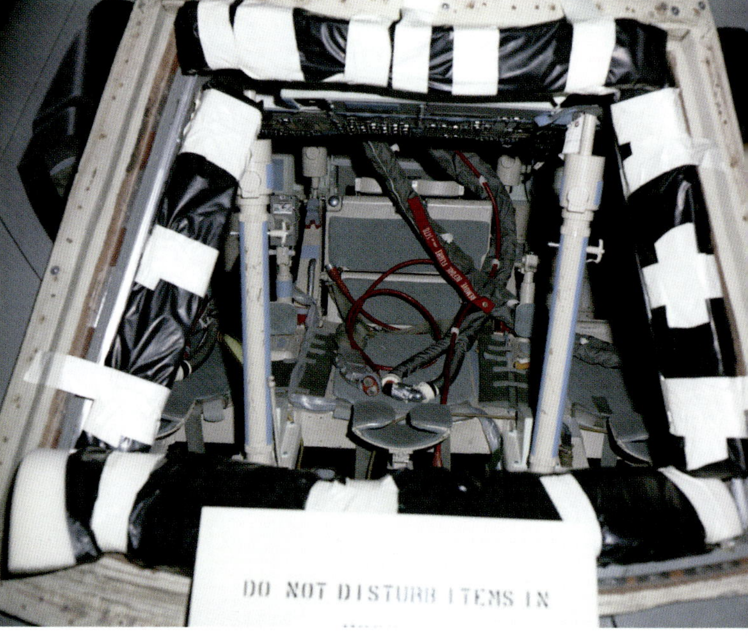

Reporters peer into CM mockup no. 2 in the PIB on March 3, the first time the news media are permitted inside. The mockup had arrived from Downey on February 8. KSC public information chief Jack King is at the top of the stairs, behind the man in the white shirt.

The mockup's switches and controls have been configured to their settings in Apollo I just before the fire.

Interior of the mockup without couches. On March 22 it is used to show the review board how the hatches could be removed.

Various components are displayed on twenty-one specially built tables in the PIB's main bay in March, with the CM in the foreground. The three crew couches and their struts are above the sign, with both hatches to the right. This area is under twenty-four-hour guard.

Closer view of the display tables on March 14. More than a thousand items have been removed when CM disassembly is completed on March 27. The CM grid pattern on the floor shows basic component locations.

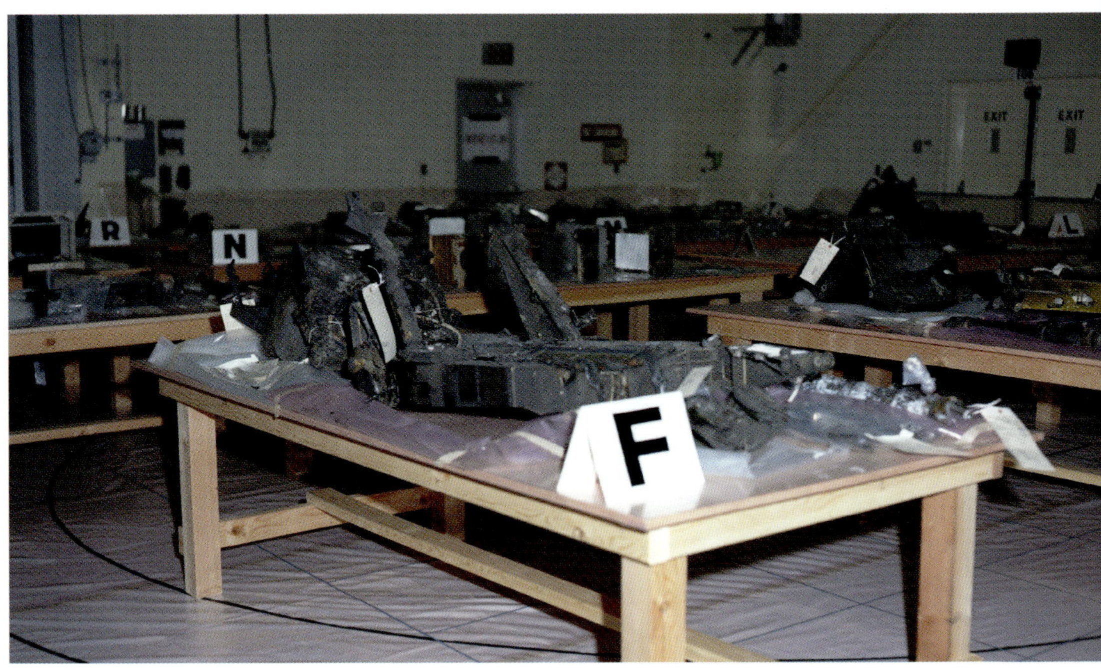

Grissom's couch is on table F, with White's couch at right.

Apollo I has been moved to a workstand on March 3.

On March 8, the CM is lifted away from its aft heat shield.

Left to right: KSC's Fred Bailey, Fire Origin and Propagation panel chairman, and board members Robert Gilruth and Max Faget examine the aft heat shield on March 8. More than a thousand MSC engineering personnel have been detailed to KSC to assist in the investigation.

The aft heat shield sits in an alignment stand in the background, with the crew compartment heat shield in the foreground on March 14.

The CM is in the final disassembly area on March 14, with the apex cover at right. A KSC security officer is at left.

The crew compartment heat shield is moved to the same area as the CM's inner shell on March 30. In early April, the aft heat shield is also moved to that location.

An investigator makes notes in the CM component bonded storage area. The doors at left lead to the work area with the CM's structures in the previous two photos.

CM-014, another Block I spacecraft, is shown in the PIB on March 29. It had been flown in from Downey on February 1 to help develop removal techniques for components inside Apollo 1. The spacecraft was originally slated for the canceled Apollo 2 mission.

CM-014 is lifted from a handling dolly inside the PIB on March 31 before it is returned to Downey.

The Instrument Unit (IU) is lowered on March 30 at the pad.

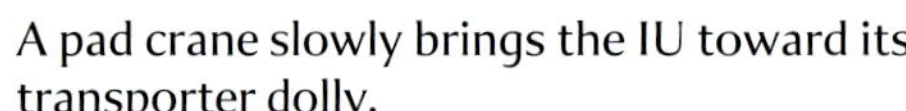

A pad crane slowly brings the IU toward its transporter dolly.

The S-IVB second stage is lowered on April 3 after demating.

The S-IVB continues down.

Pad workers carefully position the S-IVB on its trailer. Like the IU, it will be moved to nearby LC-37B and used the following January as the second stage for Apollo 5, which will put the first (unmanned) lunar module into earth orbit.

SA-204

The S-IB first stage is loaded aboard a trailer after being destacked on April 6. It would be erected the next day at LC-37B and would also be requalified for Apollo 5.

The stage is towed to LC-37, led by a KSC security escort.

The S-IB first stage is lowered to the ground as seen through the service structure.

A crane moves a fin into position during erection of SA-204 at LC-37 on April 7.

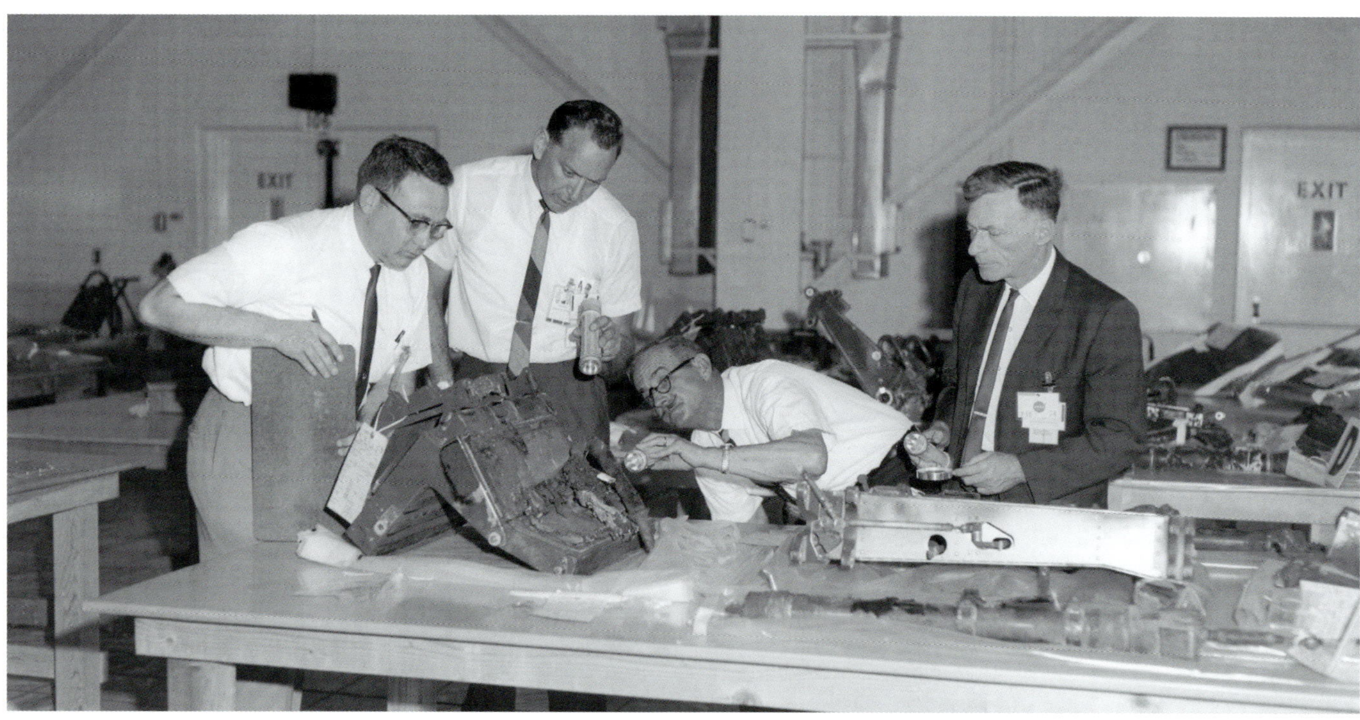

Fire Origin and Propagation panel members examine a component from Apollo I on April 9 in the bonded storage area of the PIB. *Left to right*: Al Krupnick of Marshall Space Flight Center; Robert Van Dolah, review board member and research director for the Explosive Research Center of the US Bureau of Mines; Homer Carhart; and Irving Pinkel.

Borman with Max Faget inside the Apollo CMS in the Flight Crew Training Building at KSC on April 9. The Apollo 204 Review Board's final report is made public the same day.

Review board members convene in the Management Review Center of NASA Headquarters in Washington, DC, on the evening of April 11 to review and edit transcripts of hearings the previous day before the House Subcommittee on NASA Oversight of the US House of Representatives Committee on Science and Astronautics.

With the heat shield removed, the crew compartment aft bulkhead shows a long fracture along the upper right edge where the pressure vessel failed. A hole burned in the lower right is below the ECU. The unit's glycol, used for cooling, produced an extremely intense localized fire.

The inside of the aft heat shield shows the burned area below the ECU.

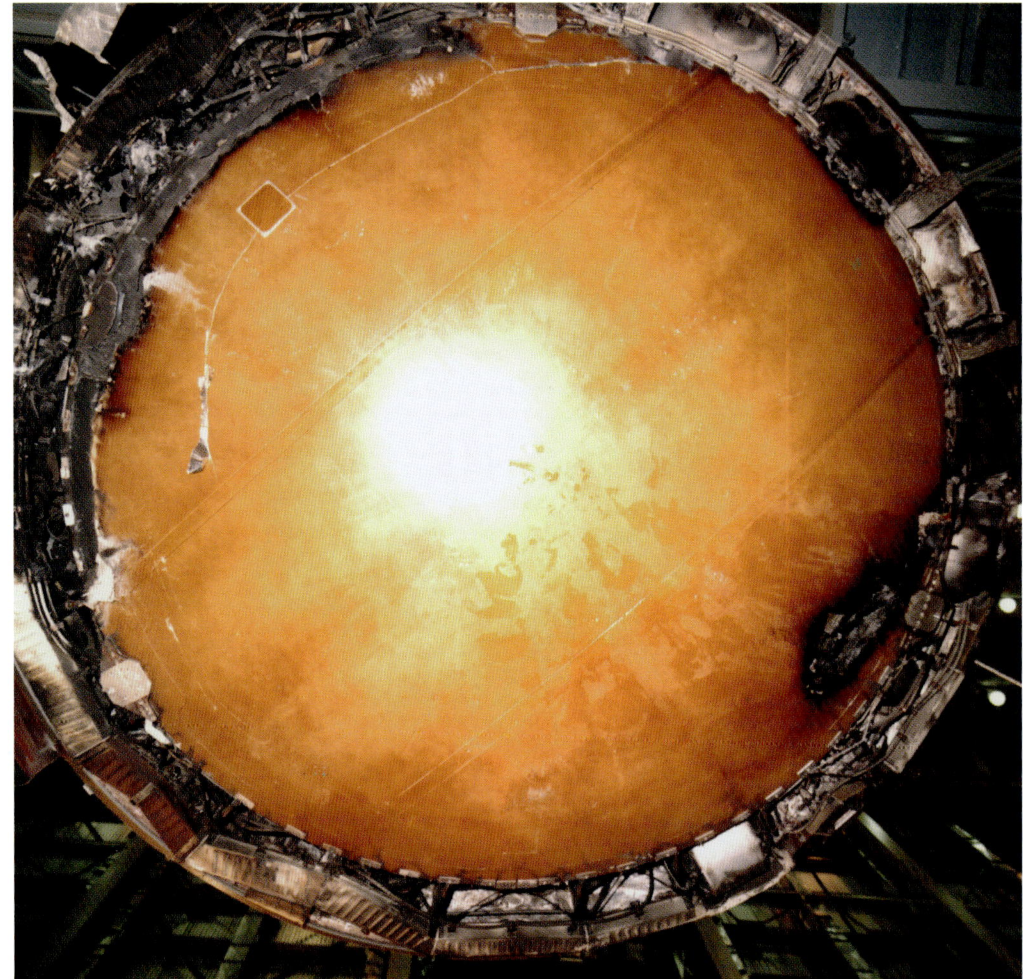

Insulation at the top of the SM below the ECU shows limited damage during inspection in the MSOB; the SM would be cleared of any role in the fire.

The SM's SPS fuel and LOX tanks underneath the insulation were unaffected.

Left to right: Astronauts Borman, McDivitt, Slayton, Schirra, and Shepard appear before the House Subcommittee on NASA Oversight in the Rayburn Building in Washington, DC, on the night of April 17, at the beginning of a second week of hearings on the accident. "I think we all got lulled into a false sense of security," says Slayton. All five agree with the review board's findings and recommendations. Posters in the back of the room explain spacecraft wiring. *UPI photo*

Borman briefs members of the House NASA Oversight Subcommittee in the PIB on April 21. They also visit LC-34 and hold a public hearing to question NASA managers and witnesses. KSC Launch Operations director Rocco Petrone (*back to camera, lower center*) weathers particularly sharp questioning. Subcommittee members are chairman Olin Teague (D-Texas), Ken Hechler (D-West Virginia), Emilio Daddario (D-Connecticut), James Fulton (R-Pennsylvania), Edward Gurney (R-Florida), John Wydler (R-New York), and Guy Vander Jagt (R-Michigan).

Deputy NASA administrator Robert Seamans, administrator James Webb, associate administrator for space flight George Mueller, and the Apollo program director, Maj. Gen. Sam Phillips, testify at a US Senate Committee on Aeronautical and Space Sciences hearing in the Old Senate Office Building on May 9. Some questions concern an internal report that Phillips had prepared in December 1965 that was critical of NAA, but Webb tells the committee the company would remain the prime contractor.

The Apollo I CM's crew compartment, its apex cover, and its heat shields are prepared for shipment to Langley Research Center in Hampton, Virginia.

The shipping container is loaded onto a barge at the LC-39 Turning Basin on the morning of July 8.

Epilogue

The Apollo 1 CM was permanently stored in this steel container in a fenced shipping-and-receiving area at Langley Research Center in Hampton, Virginia, seen in 1989. The review board designated Langley as custodian of all physical evidence, reports, files, and working documents on February 27, 1967, and the container was sent by barge from KSC on July 8, arriving at Langley five days later.

Apollo 1's forward and aft heat shields in storage in 1989, with the forward boost protective cover in place at the top. The container was originally kept in a low-pressure nitrogen atmosphere to minimize corrosion, but small leaks eventually made that impractical.

After the review board's ten-year storage mandate expired, NASA debated what to do with the Apollo I CM in early 1978, including a burial at sea, but took no action, although some records were transferred to the National Archives in Washington, DC. In April 1990, the agency moved thirty-one cartons of hardware and paperwork to Cape Canaveral AFS and announced on May 1 that the spacecraft would be placed in a deactivated Minuteman missile silo at LC-31/32, with space shuttle *Challenger* wreckage. But George Mason University junior student David Alberg, who had been researching the accident, mounted a personal campaign to stop the move, contacting several museums and astronauts. On May 31, NASA administrator Richard Truly announced that the spacecraft would not be buried but would also not be turned over for museum display, and the cartons were returned to Langley from Florida.

In 1996, Grissom's widow, Betty, requested that NASA allow display of Apollo I at the US Astronaut Hall of Fame in Titusville, Florida, for the thirtieth anniversary of the accident. The White and Chaffee families indicated that they had no objection, but NASA administrator Daniel Goldin declined, writing that "NASA has never released space artifacts related to the deaths of astronauts for exhibit."

In 1999, NASA again considered what should be done long term. Options included storage at the National Air and Space Museum in Washington, DC, or entombment in a silo, or at LC-34, or at the Grissom Museum in Mitchell, Indiana. In the end, storage at Langley was continued. On February 17, 2007, the spacecraft and remaining documents were moved to a newer, environmentally controlled warehouse 90 feet away. In 2017, Apollo I's three hatches were put on permanent display at the KSC Visitor Complex on the sixtieth anniversary of the accident.

The inner crew compartment is stored next to the heat shields. Technicians would briefly check on the contents twice a year.

Eighty-one boxes of records, components, and the spacecraft occupied about 3,300 cubic feet in the container in 1989.

The crew compartment heat shield sits outside on June 13, 1990, after being cut in half. NASA had announced on May 1 that it was planning to move Apollo 1 and the cartons to a deactivated missile silo at Cape Canaveral AFS, where debris from space shuttle *Challenger* was stored, and on May 10 the two heat shields were moved outside for the first time in twenty-three years.

The aft heat shield was also cut in half; it was 12.8 feet in diameter, but the silo was only 12 feet.

Workers return the crew compartment to the container on June 29, 1990. NASA decided on May 31 against moving it after discussions with the astronauts' families and officials at the Smithsonian's National Air & Space Museum, which had indicated it might be interested in taking custody, although not for display.

The inner hatch in the Langley storage container on March 20, 1996

The forward heat shield on March 20

Window no. 1 on Grissom's side of the cabin on March 20

The heat shield with the crew compartment in the background in a photo dated October 26, 1998

The crew compartment on October 26

Interior view of the forward and aft heat shields on October 26

Stephen Feldman, president of the Astronauts Memorial Foundation, speaks at NASA's annual Day of Remembrance ceremony on the fortieth anniversary of the Apollo I fire on January 27, 2007, at the KSC Visitor Complex. *Left to right*: Faith Johnson, daughter of late astronaut Ted Freeman; Roger Chaffee's widow, Martha; NASA associate administrator William Gerstenmaier; KSC director Bill Parsons; AMF chairman William Potter; former astronauts Walt Cunningham and John Young; and Grissom's brother Lowell. Behind them is the Space Mirror Memorial, displaying the names of astronauts who'd died during the space program. *Photo by George Shelton / NASA*

"If we hadn't had the Apollo fire, we'd have never gotten to the Moon," Young says. "Without the vehicle being built right, you'd have lost one on the way."

Ed White III touches his father's name on the mirror. Chaffee's daughter, Sheryl, is also in attendance.

Entrance to *Ad Astra Per Aspera*, the permanent Apollo I exhibit in the Apollo/Saturn V Center at KSC, dedicated on January 27, 2017, the accident's fiftieth anniversary. The area showcases personal memorabilia from the three astronauts and photos and video from their professional and personal lives. *Photo by J. L. Pickering*

The Apollo I mission emblem appears to float as it's projected at the gateway to the display area in 2018. *Photo by Mark Usciak*

The three Apollo I hatches on display in 2018. *Photo by Mark Usciak*

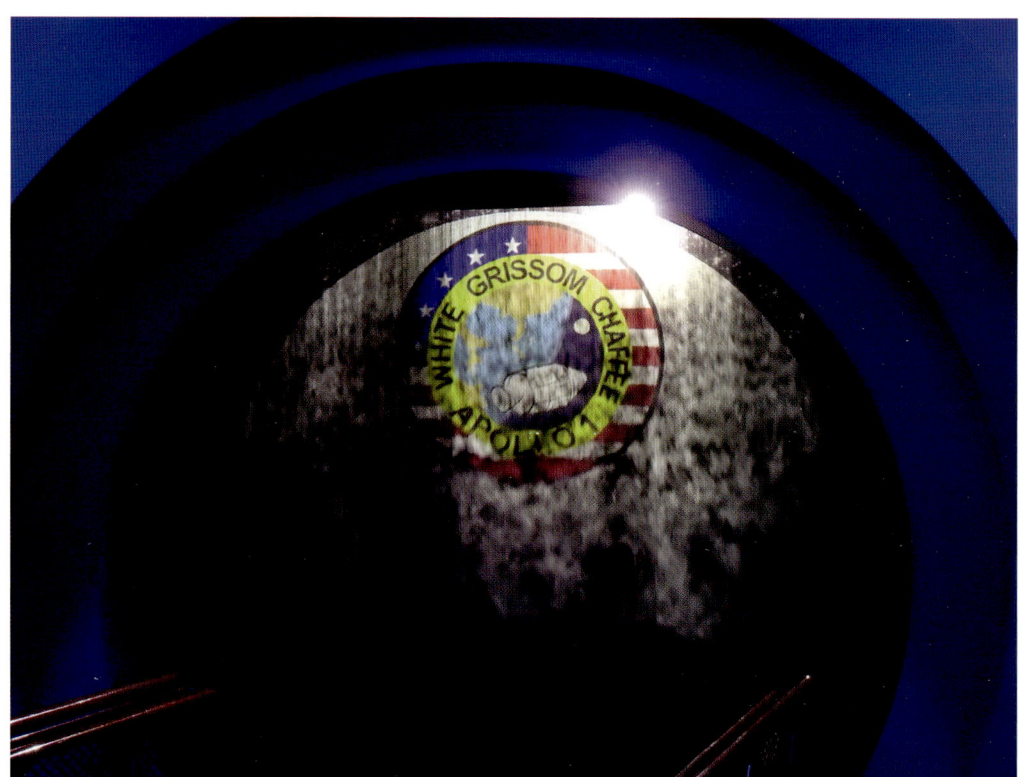

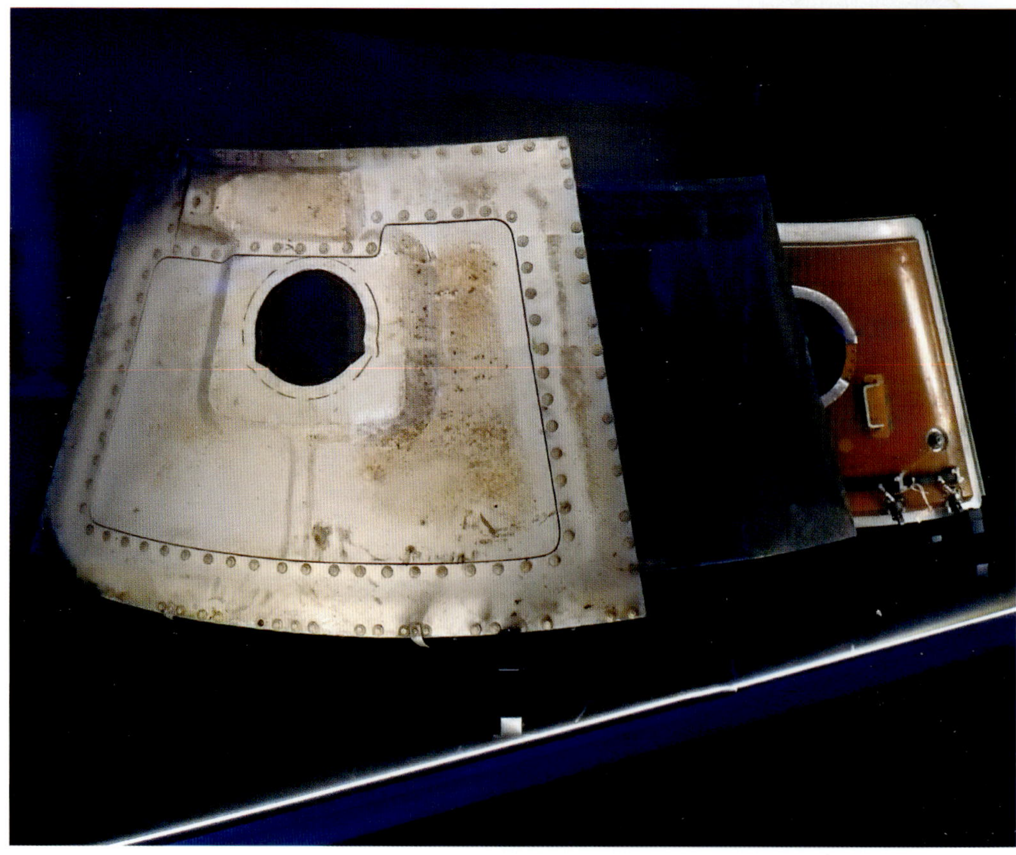

Grissom's gravestone at Arlington National Cemetery in 1978. "If we die," Grissom had said, "we want people to accept it. We are in a risky business, and we hope that if anything happens to us it will not delay the program. The conquest of space is worth the risk of life." *Photo by J. L. Pickering*

Chaffee's gravestone at Arlington in 1978. "Probably the greatest thing a man can say to himself or have as his philosophy when he has to tackle a tough job, or make a big decision, is the first eight words of the Scout Oath," Chaffee had said. "On my honor, I will do my best." *Photo by J. L. Pickering*

White's gravestone at West Point Cemetery in 2009. John White had quoted his nephew as saying, "I know there are going to be some lives lost in this program. But I feel this should not deter it. After all, aviation developers didn't quit when somebody died testing a new aircraft." *Photo by Mark Usciak*

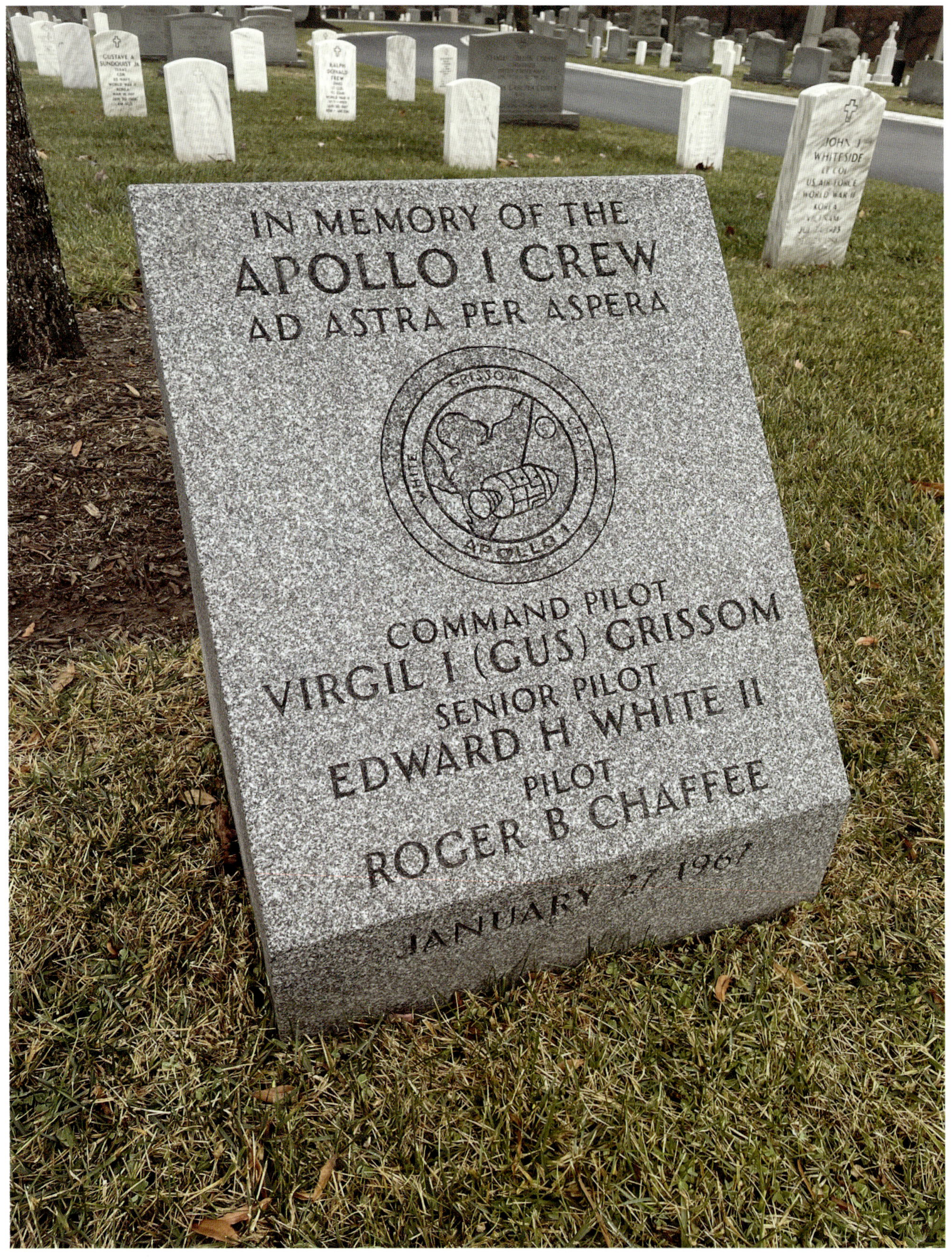

An Apollo I monument at Arlington near the Grissom and Chaffee gravesites was dedicated on June 2, 2022. The marker was funded with contributions from Aerojet Rocketdyne, Aerospace Corp., Boeing, Leidos, L3Harris, Lockheed Martin, Marotta Controls, and Northrop Grumman. *Photo by Mark Usciak*

Abbreviations

AFB	air force base
AFS	air force station
AMS	Apollo Mission Simulator
BOSU	Bioastronautics Operational Support Unit
BPC	boost protective cover
CM	command module
CSM	command/service module
ECS	environmental control system
ECU	environmental control unit
EST	Eastern Standard Time
EVA	extravehicular activity
IU	instrument unit
KSC	Kennedy Space Center
LC	launch complex
LES	launch escape system
LM	lunar module
LOX	liquid oxygen
MASTIF	Multiple Axis Space Test Inertia Facility
MR	Mercury-Redstone
MSC	Manned Spacecraft Center
MFA	Manned Flight Awareness
MSFC	Marshall Space Flight Center
MSOB	Manned Spacecraft Operations Building
MV	motor vessel
NAA	North American Aviation
NASA	National Aeronautics and Space Administration
PIB	Pyrotechnic Installation Building
RCS	reaction control system
SA	Saturn
SLA	Spacecraft / Launch Vehicle Adapter
SM	service module
SPS	service propulsion system
US	United States
USAF	United States Air Force
USN	United States Navy
VAB	Vehicle Assembly Building

Bibliography

Alexander, George. "Inquiry Focuses on Electrical Systems." *Aviation Week & Space Technology*, February 6, 1967.

Benson, Charles D. *Moonport: A History of Apollo Launch Facilities and Operations*. Washington, DC: Scientific and Technical Information Office, National Aeronautics and Space Administration, 1978.

Blackburn, Gerald. "Life at 'The Spaceship Factory.'" *Palos Verdes Pulse*, June 27, 2020. Palos Verdes Pulse, Palos Verdes Peninsula, CA.

Brooks, Courtney G., and Ivan D. Ertel. *The Apollo Spacecraft: A Chronology*. Vol. III. Washington, DC: National Aeronautics and Space Administration, 1976.

Chrysler Corp. Space Division. *Saturn IB Orientation Systems Training Manual 851-0*. Huntsville, AL: Chrysler Corp. Space Division, Huntsville Operations, Engineering Communications Dept., 1965.

Chrysler Corp. Space Division, Huntsville Operations, Engineering Communications Dept. *SA-9 Vehicle and Launch Complex Functional Description—Launch Pad Accessories*. HEC-D042, vol. VII. Huntsville, AL: Chrysler Corp. Space Division, Huntsville Operations, Engineering Communications Dept., 1964.

Coan, Paul P. "Apollo Experience Report—Television System." NASA Technical Note D-7476. Houston: National Aeronautics and Space Administration, Lyndon B. Johnson Space Center, 1973.

Deming, Joan, and Patricia Slovinac. *Historical Survey and Evaluation of Facility 49635 / Environmental Health / Health Physics Facility, (BOSU), Cape Canaveral Air Force Station, Brevard County, Florida*. Sarasota, FL: Archaeological Consultants, 2012.

Department of the Interior, National Park Service, Southeast Region. "Cape Canaveral Air Force Station, Launch Complex 39, Altitude Chambers." Atlanta: Historic American Engineering Record, 2009.

Ertel, Ivan D., and Mary Louise Morse. *The Apollo Spacecraft: A Chronology*. Vol. I. Washington, DC: National Aeronautics and Space Administration, 1969.

General Motors Corp., AC Electronics Division. *Apollo Guidance and Navigation System Study Guide—Block I (Series 100) G&N System Familiarization*. Milwaukee, WI: General Motors Corp., AC Electronics Division, 1965. Revision B, 1966.

Knacke, T. W. "The Apollo Parachute Landing System." Paper presented at the AIAA second Aerodynamic Decelerator Systems Conference by Northrop Ventura, El Centro, CA, September 1968.

Kozloski, Lillian D. *U.S. Space Gear: Outfitting the Astronaut*. Washington, DC: Smithsonian Institution Press, 1994.

Morse, Mary Louise, and Jean Kernahan Bays. *The Apollo Spacecraft: A Chronology*. Vol. II. Washington, DC: National Aeronautics and Space Administration, 1973.

National Aeronautics and Space Administration. *Apollo Recovery Operational Procedures Manual, Revision C*. Houston: Manned Spacecraft Center, Landing and Recovery Division, 1971.

National Aeronautics and Space Administration. "Crew Egress Procedures for Apollo Block 1 Command Module at Sea." NASA Program Apollo Working Paper 1213. Houston: Manned Spacecraft Center, 1966.

National Aeronautics and Space Administration. "Launch Complex 34 Facilities." KSC Fact Sheet 05, 1968.

National Aeronautics and Space Administration, Launch Operations Center. "NASA Space Utilization at AMR." Cape Canaveral, FL, 1962.

National Aeronautics and Space Administration. *Report of Apollo 204 Review Board to the Administrator, National Aeronautics and Space Administration*. Washington, DC, 1967.

National Aeronautics and Space Administration (Marshall Space Flight Center, Kennedy Space Center), Chrysler Corp. Space Division, McDonnell Douglas Astronautics Company, IBM Federal Systems Division, and Rocketdyne. *Saturn IB News Reference*. 1965.

New South Associates. *Cape Canaveral Air Force Station, Launch Complex 39, HAER No. FL 8-11-T-3, Hypergol Maintenance and Checkout Area, Hypergol Support Building*. Atlanta: Historic American Engineering Record, National Park Service Southeast Region, Department of the Interior, 2013.

North American Aviation, Space and Information Systems Division. *Apollo Operations Handbook—Command and Service Module—Spacecraft 012*. Downey, CA: North American Aviation, Space and Information Systems Division, 1966.

North American Rockwell. *Apollo Spacecraft Familiarization Manual*. Downey, CA: North American Rockwell, 1967.

North American Rockwell, Space Division. *Apollo Spacecraft News Reference.* Downey, CA: North American Rockwell, Space Division, 1969.

Pavlosky, James E., and Leskie G. St. Leger. *Apollo Experience Report Thermal Protection Subsystem.* Houston: National Aeronautics and Space Administration, Lyndon B. Johnson Space Center, 1974.

Prentice, R. W. "Transportation of Douglas Saturn S-IVB Stages." Paper presented to the American Society of Civil Engineers by Douglas Missile & Space Systems Division, Cocoa Beach, FL, November 1965.

Slovinac, Patricia. *Cape Canaveral Air Force Station, Launch Complex 39, HAER No. FL-8-11-E Altitude Chambers.* Atlanta: Historic American Engineering Record, National Park Service Southeast Region, Department of the Interior, 2009.

Tylko, John. "Simulating Apollo: Flight Simulation Technology, 1945–1975." PhD diss., Massachusetts Institute of Technology, 2023.

Index

Note: the following entries are too numerous to include: Apollo 1; Chaffee, Roger; command module; Grissom, Gus; Saturn IB; S-II, S-IVB, service module; White II, Ed.

GRISSOM
WHITE
CHAFFEE
APOLLO 1